Secrets *of* Stargazing

SKYWATCHING TIPS AND TRICKS

For my parents, who allowed me to stay out late when I was young. And for Shane, who stays out late with me now.

OTHER BOOKS IN THIS SERIES

Patterns in the Sky by Ken Hewitt-White
Exploring the Moon by Gary Seronik

Sky Publishing
90 Sherman Street
Cambridge, MA 02140-3264, USA
SkyTonight.com

Library of Congress Cataloging-in-Publication Data

Ramotowski, Becky.
Secrets of stargazing : skywatching tips and tricks / Becky Ramotowski.
p. cm. -- (Astronomy for everyone series)
Includes bibliographical references and index.
ISBN 978-1-931559-40-9
1. Stars--Observers' manuals. 2. Astronomy--Observers' manuals.
3. Astronomy--Amateurs' manuals. I. Title.
QB63.R323 2007
522--dc22
2007002799

Printed in China

Secrets *of* Stargazing

SKYWATCHING TIPS AND TRICKS

Becky Ramotowski

SKY PUBLISHING
A New Track Media Company
Cambridge, Massachusetts

Table *of* Contents

Welcome *to a Lifestyle*

I wish I'd had an astronomy muse when I was growing up. Oh, I didn't need to be inspired to look at the sky; I always seemed to be doing that anyway. No, what I could have used was someone passing on morsels of accumulated wisdom to help speed my transition from eager novice to enthusiastic, knowledgeable amateur astronomer. While I don't claim to know everything about stargazing, I've learned an awful lot since I started looking up. But it took a long time, and it was sometimes harder than it needed to be. That's why I wrote this book: to simplify and speed up the learning process for anyone interested in exploring the night sky.

Secrets of Stargazing is full of quick and easy tips that will ease you into, and through, many nights of observing without explaining, in excruciating and unnecessary detail, how everything works. (That's what manuals are for; this isn't one.) I've skipped the tedious instructions and offer only helpful tricks and useful shortcuts so you can spend more time observing and less time fumbling and grumbling in the dark. I'll tell you how to find the darkest observing site possible no matter where you live, how to easily (and inexpensively) improve your skywatching experiences with and without a telescope, and let you in on a few of the secrets I've collected from my many nights under the stars.

Some have called me passionate about astronomy. It could be because I once temporarily left my mother-in-law's birthday party so I could watch and photograph the Moon covering and uncovering Saturn. Or perhaps it's because of the time I almost fell into a lake while looking up instead of watching where my feet were going. Or maybe it's because I was once in such a hurry to observe that I didn't give my eyes time to adjust to the dark, raced outside, and smashed face first into a dumpster. These are tales for another time, though the secret they reveal is that amateur astronomy can sometimes be hazardous!

The sky has been my playground for as long as I can remember. Yet I still consider myself a student of the sky because there's always some unfamiliar celestial territory out there to explore. For me, astronomy isn't just a weekend hobby — it's a lifestyle. Over the years I've observed hundreds of objects from many locations (I keep a sky diary, and you should too). This routine has provided me with many valuable insights into how to make observing a better experience — wisdom that I'm happy to share as *your* astronomy muse.

It's no secret that anyone can go outside and simply look up, but I think I've gathered together some sure-fire tips and techniques that'll quickly improve your stargazing experience. That's what *Secrets of Stargazing* is all about.

The night sky is usually no match for urban lighting.
Mars is barely visible to the right of the tower.

From the Balcony *to the Back Roads*

When I was a young astronomer growing up in rural East Texas, my night sky was a beautiful blanket of stars that kept my inquisitive eyes occupied every clear night. The Milky Way was a mist of precious stellar jewels, and shooting stars left long, lingering trails on the ebony background overhead. At times it seemed as if the stars were glowing just a few country miles above the tall pines in my parents' yard and not shining inconceivably far out in space. It was a wondrous time — one whose memories I'll always cherish.

Sadly, my entry into the working world required a move to a sprawling metropolis of 2 million inhabitants, where it seemed that the entire population never slept and was totally light dependent. Still, I was determined to observe — even if I could see only a handful of stars.

A popular saying, "When the going gets tough, the tough get going" nicely summarizes the task of the urban astronomer. There are many ways to get tough and overcome some of the obstacles found in less-than-ideal observing locations without packing up the minivan and leaving town, though that's also an option if all else fails.

No telescope? No problem. Public star parties are often held on a regular basis by local astronomy clubs — even in New York City.

Adapting to Apartments

If you call an apartment complex home, just a few easy changes can improve your observing environment considerably. Shielding porch or balcony lights, turning off interior room lamps, and closing the curtains may be all that's needed to allow a more enjoyable night of viewing from a small patio or balcony. Don't forget to switch off the television and computer monitor. Consider replacing a few white-light bulbs with red ones in lamps by the door to help preserve your dark adaptation (see page 57) if you find yourself going in and out a lot. And if the refrigerator is by the back door, think about temporarily replacing its bulb, too. Just warn your family that the ketchup will disappear in red light!

If you live in an apartment complex and can get permission to observe from its roof, you may find the perfect skygazing site — given that your location is in the city.

Encouraging your neighbors to turn off unnecessary outside lights and close their blinds will help eliminate other unwelcome light and glare. In return, be a good neighbor and invite them over for a peek through your telescope as a thank-you gesture. The first time they gaze at the Moon or Saturn in your scope should encourage them to keep their lights down whenever they see you outside with your gear. It'll also reassure them that you're not some weirdo or Peeping Tom.

While it's not always possible or practical, think about observing after 11 p.m. Many neighborhood stores are closed by then — some even turn off their lights — thereby reducing, at least a bit, the amount of skyglow filling the heavens.

An apartment balcony might be a cramped viewing site, but it can offer a high platform that provides a fine, uninterrupted view of the sky from zenith to horizon.

BEYOND THE BALCONY

If stargazing from your apartment balcony or patio isn't an option, look for other on-site locations. Courtyards, playgrounds, or picnic areas set aside for tenants may be available for use after hours (don't forget to ask permission) and might turn out to be perfect urban skygazing spots. Avoid wooden decks if you can; they tend to vibrate and don't make good observing platforms.

Fences, boundary walls, or tall trees and shrubs can be used to block stray light and that chilly breeze. The key is to be creative and not discount anything as a possible light stopper. You need to be a thorough detective and investigate every nook and cranny of your facility as a potential parking space for a telescope. Look around in daylight, and then check it again at night before hauling out your gear. This extra effort will pay off come observing time.

At one of my former apartment-complex homes, I observed from behind a very large dumpster. Fortunately it was one for discarded furniture and small appliances and was virtually odor- and varmint-free!

URBAN OBSERVING TARGETS

So what can you see from your urban observing site? Start by selecting targets that make viewing less like work and more like play. The Moon should be at the top of your list. It's easy to find, is in the evening sky for roughly half of every month, and has a wealth of fascinating features that can keep you occupied for years. Watching the nightly (or even hourly) changes along the lunar *terminator* — the line that divides day from night — will easily turn minutes into hours of satisfying viewing.

The bright planets — Venus, Jupiter, Saturn, and sometimes Mars — are also easily seen from an urban environment. Dim little Mercury can occasionally be spotted, particularly if you're able to use the Moon or Venus as a directional marker. Even Uranus and Neptune may be glimpsed from an urban setting if seeing conditions are good, your light pollution isn't too severe, and you know exactly where to look.

It may seem obvious, but don't forget the stars! Double stars make fun targets — many are easy sights in a small scope. Bright open star clusters such as the Pleiades and Hyades, and globular clusters like the Great Cluster in Hercules, should also be on your list of objects to hunt. It's also possible to view nebulas and galaxies from a city, though if you're a novice stargazer, all but a handful of these will be difficult to spot. (I've included my favorite urban objects in Appendix 2.)

Finally, keep an eye out for those occasional events that don't need dark skies to be impressive. Lunar eclipses, the best meteor showers, conjunctions (gatherings) of planets, the appearance of a bright star or planet next to the Moon, the passage of the International Space Station — all can be easily seen from an urban environment.

Moon and Mars

Saturn

Moon and Pleiades

Backyard Bliss

A backyard is a wonderful thing. Homeowners in an urban setting with such a chunk of flexible real estate at their disposal have a significant advantage over cramped apartment dwellers. Here you can set up everything from a lawn chair for comfortably cruising the sky with binoculars to a telescope with all its trappings. A backyard provides more security than the open spaces in an apartment complex, and there's the option of being able to leave equipment outside all night if needed.

Of course, even a backyard isn't immune to light-pollution problems. Getting to know the neighbors remains a good idea, since their outside lights will affect your viewing. Host a neighborhood starwatching session occasionally so you can demonstrate the benefits of keeping extraneous light to a minimum. But even if you can't find a light-free corner of the yard, you can set up, and leave in place, semi-permanent light shields — something as simple as stretching a piece of dark cloth or a tarp between two poles *(left)* or propping up a large patio umbrella.

In a perfect universe, every amateur astronomer would have a personal observatory in the backyard, where everything is set up and ready to go. Of course it doesn't have to be a dome — in urban areas where building codes rule, an observing shed is often a better solution. For most beginners a personal observatory is just a dream, but as you gain observing experience, don't discount the possibility.

URBAN OBSERVING TIPS

Cover your head with a dark cloth, hood, or towel while looking through the eyepiece (see page 28 for more about eyepieces). Okay, it looks pretty strange, but it does help block stray light and nobody's going to see you anyway.

Wear an eye patch over your observing eye when it's not looking through an eyepiece. This will help keep your eye adapted to the dark.

If you must duck into the house (and haven't installed red-light bulbs everywhere), **wear red-tinted glasses or goggles** while inside. These will also help protect your night vision.

Special **eyepiece filters** will help some dimmer objects "pop" out of the murk of a city sky. (There's more about filters starting on page 31.)

Farther Afield: City Parks

Sometimes your regular urban observing site isn't good enough. Perhaps the very trees that shield some lights also block your view of an upcoming astronomical event. Or maybe you want to help organize a group observing session to catch an eclipse, conjunction, or other astronomical sight. In such cases, your driveway, backyard, or cramped apartment balcony/patio simply won't do.

Don't just hop in your car and drive aimlessly around unfamiliar neighborhoods hoping to stumble across a good site. Start by examining a city map for some idea of where the best prospects might be located. Recreational areas such as baseball and soccer fields are usually marked on detailed maps; city parks should be obvious. With a little sleuthing, you can probably discover a flat, open area that'll provide a roomier locale.

Next, enlist the help of an observing buddy and check these sites during the day. Can you park beside the field, or will you have to carry your equipment to a suitable spot? Is it waist-deep in snow? Are there too many trees and not enough open space? Is there a handy bench or picnic table where you can set up? Then go back at night, preferably around the time you want to observe. Are sprinklers watering the field? Are there too many lights, or lights that happen to be in exactly the wrong location? If you've chosen a recreational area, is there a game going on? A few phone calls to local officials should provide answers to these questions.

If there are just two of you heading out (and I strongly recommend against solo observing), you should be all set. But if you're part of a larger group, make sure you've contacted park security or the local authorities with jurisdiction in the area to tell them of your plans. Remember, city parks are public places. Anyone may stop and want to know what's going on and what you're looking at. Most people are unaware of nightly celestial events, and even if there is nothing special happening, they'll probably enjoy a peek through your scope.

Always keep your personal safety in mind. Patrolling officers may provide some protection (and be interested observers themselves), but if anything happens that makes you feel uncomfortable, pack up and leave immediately.

Get Outta Town!

The star clouds of the Milky Way, invisible in an urban setting, sparkle when viewed from a rural site.

If you live in a city but yearn for a dark sky, you're going to have to travel. Fortunately, just an hour or so of driving time may be the difference between an illuminated urban balcony and a reasonably dark rural observing site. The drive could result in magnitudes of improvement when it comes to seeing dim fuzzies through your binoculars or telescope.

To find a dark back-road viewing haven, use some of the techniques you employed to hunt down a city-park observing site. Again, a good map is the key. If you have, or can borrow, a GPS (Global Positioning System) unit, use it as you navigate your way to possible observing sites. Finding them again will be much easier, particularly if some of the roads are not well traveled or marked with signs.

Be prepared before venturing out long distances (whether site searching or actually observing). Most remote rural areas don't have 24-hour convenience stores, so a full tank of gas, a small emergency kit, and a pack with a few snacks and beverages should be part of your observing kit. Toss in a couple of blankets, too, just in case it gets chilly. Carry your cell phone and tell family members or friends where you're going — or take them along for company.

Where to observe? Rural farms and ranches are one option. Don't trespass; be sure to ask permission of the landowner, and make sure there are no glaring all-night farmyard lights nearby. You could discover that the owner likes to stargaze, maybe wishes he/she owned a telescope, and is happy to have you drop by occasionally and share the sky.

Public lands are another option. These are areas set aside for various recreational activities and are sometimes free or accessible for a small fee. Some have campsites with overhead shelters and picnic tables, running water, and restroom facilities. These ameni-

SHARING YOUR SPACE

Remember: you're probably not alone at your dark-sky site. It could be the abode of many wild animals that go bump or buzz (or growl) in the night. Who knows what critters are lurking in those woods over there! This is their home, so try not to startle them or make them feel threatened. And don't attempt to touch or feed them no matter how cute or friendly they seem.

Every region has its own particular brand of two-, four-, or even no-footed creatures to watch out for, so I won't list them here. But various species of small (often invisible) biting beasties are ever-present across much of the Northern Hemisphere, so be prepared. And watch out for prickly plants, too. The unfamiliar territory you've selected as your viewing site may be the habitat of some less-than-cordial flora.

If possible, keep your vehicle parked nearby with the doors unlocked and the windows up just in case a hasty retreat from an animal is required. And keep your cell phone physically at hand, not stuffed into one of your equipment bags. Finally, if possible, always observe with a buddy.

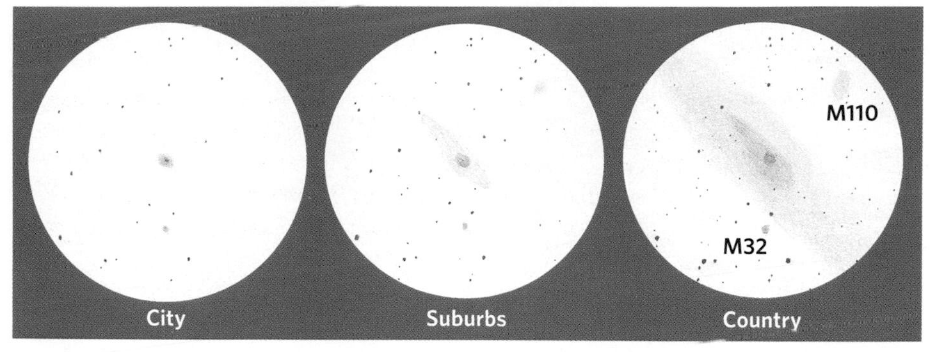

Here's graphic, visual proof why you should get out of town to observe. These sketches show how the Andromeda Galaxy (M31) appears through a 2.8-inch (70-mm) telescope at 30× from an urban park *(left)*, in the suburbs *(center)*, and under dark country skies *(right)*.

ties make them ideal locales for comfortable all-night observing for families or groups of astronomers wanting to get away from light-polluted city skies.

Rural ballparks and soccer fields should also be looked at as potential observing sites. Some are very rustic and only offer a flat parking lot or weedy field from which to observe, but so what? Many lack lights for night games and hence are a good spot to set up for a great evening of viewing with friends.

State or provincial parks (or historical sites) are another option. Some actually encourage astronomers to set up, but beware — you may find yourself hosting a small public observing session in exchange for some dark-sky viewing later on. Any potential park should be visited in advance to determine whether it's suitable for use and what type of facilities or lighting problems may exist, and, most important of all, to obtain permission from park officials. Park gates are often locked late at night; you wouldn't want to get trapped inside.

Be careful when setting up at a rural site. Don't needlessly trample the ground cover, and watch out for rocks, stumps, or very uneven ground — anything you might trip over in the dark. Don't forget to take away everything you bring, including garbage.

The Scoop on Scopes

Congratulations! You're finally the proud owner of a new telescope. (Or perhaps you've just rediscovered an old one in the back of your closet and decided to give it another try.) I'll bet you can hardly wait to take it outside to explore the wonders of the universe.

Do yourself an enormous favor. Wait. Setting up and operating a telescope can be very intimidating the first few times you try it. Doing it in the dark just makes matters worse. Before you take the telescope outside for its official "first light" (your first look through it at a celestial object), familiarize yourself with the instrument, how it fits together, and how it operates — and do so inside, during daylight, at your own speed.

Get Acquainted

When you first unpack the box the telescope comes in, you may wonder what all that stuff is, how all those bits fit together, and how everything works. The answers to these and other questions can be found in a logical place: the owner's manual. (If it's not in the box, it's likely on the company's website.) Manual writers know you're not likely to read the whole thing, so they usually include something like a "Quick Start" or "Getting Started" section. If you read nothing else . . . read this!

Of course you're anxious to get outside and observe. But if you're unpacking your new telescope for the first time, don't do it outside as darkness falls. Do it indoors where you can see what you're doing and are less likely to lose components if you drop them.

Instead of fumbling in the dark with unfamiliar bits and pieces, assemble your scope inside. Make sure all the parts are included before you start putting it together; if everything you need isn't there, contact the manufacturer. A couple of indoor daytime practice sessions will make outdoor assembly in the dark less stressful and time consuming. In other words, make your mistakes before it matters. If you want a good "final" simulation, turn on your red-light flashlight, turn off the room lights, and put the telescope together one last time — in the dark.

If your scope has a motor drive (and perhaps a computer), install the required batteries and make sure everything works. Does the scope need a power cord or other cables? Now's the time to figure out how best to arrange them so they don't trip you in the dark or wind around the scope as it moves. If you have a Go To telescope (see page 16), power up and work your way through the setup. It doesn't matter if you're inside and can't point at Polaris or any other star — fake it. The computer doesn't care, and you'll be able to see how all the controls function and how long the various motions take.

At the end of your practice session and before you disassemble the instrument, consider where you're going to store it and how much of it can remain together. If it needs to travel only a short distance — from its storage nook to a balcony or backyard — perhaps the scope can stay in two or three major pieces and not have to be reassembled from scratch every time you use it.

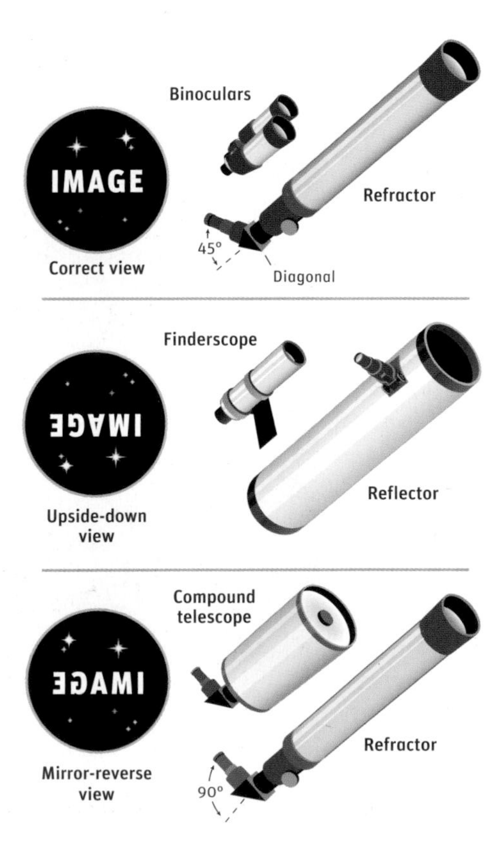

WHICH WAY IS UP?

Confusion often reigns when a novice scope user stares into the eyepiece, adjusts the telescope, and discovers that the familiar directions of up, down, left, and right sometimes don't apply. The design of the scope and the type of star diagonal it uses (if any) determine directions. Binoculars and refractors with 45° diagonals show a "normal" sky — the one you see without optical aid. Reflectors and straight-through finderscopes (those without a 90° star diagonal) turn the image upside down. When a 90° diagonal is used (on refractors or compound scopes), the image is right-side up but mirror-reversed (that is, flipped left-right).

TAKING IT OUTSIDE

There's nothing worse than attempting something for the first time — for real — in front of an audience. So you might consider trying out your new scope by yourself. The exception would be if you have an amateur-astronomer friend or neighbor; in this instance that person's presence and expertise may be very welcome. Also, restrain pets and make sure toddlers aren't around. Small, shiny telescope parts are very tempting and can easily be swallowed or carried off, never to be seen again.

If you're assembling your telescope outside (maybe you've traveled to a dark site and needed to disassemble the scope to get it in the car), put a tarp or a scrap of outdoor carpet on the ground underneath it. This will make dropped pieces easier to find and may even prevent breakage if an eyepiece or filter slips from your hands. If you're in an arid environment, covering the ground has the added benefit of reducing the amount of dust that's kicked up around the base of your scope as you observe. Regardless of where you set up, make sure your site is as level as possible.

Finally, there's no right or wrong way to arrange your observing area. I've watched many astronomers organize their scopes, tables, and chairs in a multitude of ways and am reminded of various species of birds making nests. Everyone's different — feel free to experiment. Eventually you'll find the ideal place for everything . . . at least until you buy another piece of gear!

KNOW YOUR SCOPE

A bewildering variety of telescopes inhabit today's marketplace, and offering advice about choosing the best one is beyond the range of this book. But if you're just starting out in astronomy and happen to own a telescope, it's important to know exactly what you have.

A refractor has a lens at the front end of a tube that gathers light and brings it to focus in an eyepiece at the opposite end. Refractors are basically maintenance-free but can be expensive to purchase.

A reflector has a mirror at the bottom of a tube that bounces incoming light toward the top of the tube, where a small, flat secondary mirror angles the light off to one side, to the eyepiece. Reflectors (also called Newtonians or Dobsonians) are popular because a large one can be had for a reasonable price.

MOUNT UP

The best telescope for you is the one you'll use the most. If you're a beginner and you have a choice, do not start out with a scope that has a German equatorial mount. It's the most complex to set up and use, and unless you're diving into astrophotography, you don't need it. Regardless of the mount type, make sure it's solid. A wobbly mount can make the best optics in the world perform poorly.

The altazimuth (altaz) mount has two motions — altitude (up and down) and azimuth (side to side) — and is primarily used with small refractors.

A Dobsonian is an altazimuth mount that's well suited to large reflectors. A Dob isn't motorized (though tracking platforms are available).

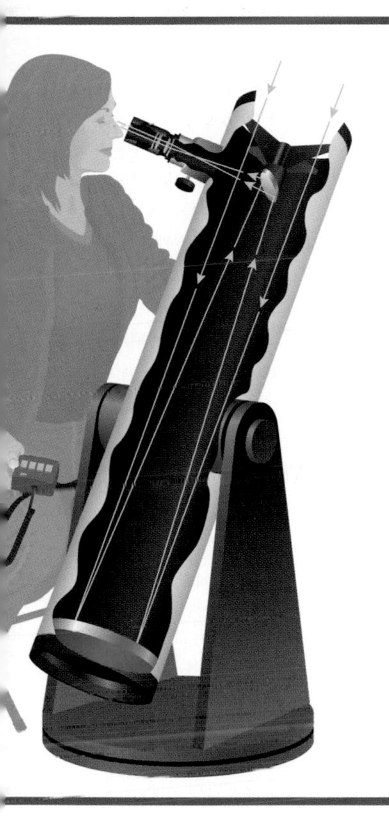

A compound, or catadioptric, telescope is a hybrid — it uses a lens (corrector plate) and two mirrors to bring the incoming light to focus at the back of the scope. Commonly referred to as "cats" or SCT's (Schmidt-Cassegrain telescopes), these scopes are compact, portable, and well suited for computer control, but they can get expensive.

An equatorial mount is designed to easily compensate for Earth's rotation — it turns on a single axis so tracking the stars becomes effortless. It's sometimes called a GEM (German equatorial mount), is used by reflectors and refractors, and can be motorized.

A single- or dual-fork mount is standard with a compound telescope; a smaller scope usually has the single fork. This type of mount (technically an altaz) is motorized and may have an onboard computer that turns it into a Go To mount (turn the page).

KEEP IT TOGETHER

An easy way to keep track of all of your telescope stuff is to set aside an area in your house or garage for all its parts and accessories. If your telescope came in a box, use the box at least temporarily for storage. But eventually you'll want to acquire a dedicated case for extended storage and to protect your investment. Small scopes on simple tripods can be left assembled and covered for easy access. Larger units that can break down into just two or three major components can also be set aside and covered. An inexpensive plastic storage bin makes a great container for keeping all the small bits together, including any tools you need. But don't let your hardware mix with eyepieces or filters; keep your optics in separate containers within the larger bin.

If you keep your gear in a garage or shed and observe on or near pavement, a wheeled tool chest makes a great storage unit. Some amateurs go whole hog, as illustrated by this custom-built, wheeled desk-cart and storage table — an all-in-one rolling observatory kit!

"Go To" Delights

This book isn't about how to buy a telescope, but if you don't have one yet (and want one), I know that the possibility of acquiring a Go To scope will cross your mind. So here are a few ideas and tips to help you better understand this very popular technology. (And if you have a Go To that's not properly "going to," perhaps I can help.)

Even with a Go To, you still need to know something about the sky. In the image above, which star is which? (Hint: This is part of the winter sky.) Aligning the telescope on the wrong bright star will cause the scope's electronic brain to become lost.

A Go To is a motorized telescope equipped with a small computer (located in either the base of the scope or its mount) that contains a database loaded with thousands of objects. With the push of a few buttons on a hand controller you send the telescope toward any object in that database, look into the eyepiece (after the scope stops moving), and see the celestial sight. At least that's the theory.

Many novices find this type of telescope very appealing because it lets them spend more time observing and less time searching. In particular, a Go To removes the frustration of trying to locate deep-sky sights in a light-polluted sky. (Of course light pollution does overpower faint objects, so a dim nebula or galaxy may still be invisible even if the scope is pointing directly at it.) And in a dark, Milky-Way-filled sky, a Go To will help a novice observer discover dozens of deep-sky splendors beyond the obvious ones. These computerized marvels have definitely changed the way telescope users spend their time under the stars.

IT'S ALMOST MAGIC

I'll admit that some of the expensive models of Go To scopes are the closest thing to telescopic magic that I've ever seen. Some have a built-in GPS that uses signals from satellites to figure out their location. Others need to identify only one star in order to be ready to go — all the user has to do is center the star in the eyepiece to fine-tune the alignment. Also available are reasonably priced, computerized scopeless mounts, which means you can turn your favorite "old" telescope into a Go To marvel.

SET UP — GET IT RIGHT

As advanced as a Go To telescope may be, it's not magic, though the expensive models come pretty close! You can't just plunk a Go To outside, push a few buttons, and expect it to instantly find the Orion Nebula or even the Moon. It needs to be properly set up. If you've got a Go To and it's not going where you want it to, it's more than likely that your poor little scope doesn't know where it is.

Start by assembling your scope indoors during the daytime. (Sound familiar?) Read the couple of pages in the manual that describe the setup procedure (this can vary by model and manufacturer), then power up. Make sure you enter your location correctly — extreme precision for your latitude and longitude isn't necessary, but ensure that your time-zone and daylight-saving-time settings are correct. You need do this only once. If you've already entered the data, call it up and double-check that it's correct. Then pretend you're outside and follow through to the end of the Go To's setup routine. Be certain you understand what the computer is asking (you'll see data and queries displayed on the little screen in the hand controller) and know the answers before you step outside to do the setup in darkness.

Some of the responses you provide require that you have at least a passing familiarity with the stars and constellations. It's *not* true to

PUSH TO

It may seem as though Dobsonians have been left behind in the Go To revolution, but not so. Some Dobs are now available with computerized object-finding capability. You use the hand paddle to tell the scope what you want to see, and it tells you which way to push the tube (hence the nickname "Push To"). Eventually the paddle's display goes to zero, you look into the eyepiece, and the object should be there. As with a regular Dobsonian, a Push To doesn't have a motor, so you also have to manually push/pull the scope to keep the object in the field of view. The only battery required is for the hand controller, but if it fails you still have a perfectly serviceable telescope. Push To's represent a practical compromise between high- and low-tech instruments.

say that the owner of a Go To doesn't need to know the sky. Even directional knowledge is important — realizing the scope is pointing south but the object it's looking for is in the west will speed your trouble-shooting.

You communicate with your Go To by using a numeric keypad to enter your location, answer questions posed by the computer, and indicate the objects you want to view.

THINGS CAN GO WRONG

Once you're set up, beware of things that go bump in the night. In this case it's you or your observing buddies. If your Go To has been performing flawlessly but suddenly starts missing its targets, chances are the tripod has been kicked or the scope's been nudged. The only solution is to repeat the setup procedure. And if the telescope seems to have lost its mind — it's moving erratically, stalling, or otherwise misbehaving — maybe it has. Remember the Go To is actually a computer, and computers sometimes crash. Try powering down and then up again. If this doesn't solve the problem, or if the hand controller's screen shows gibberish, or if the scope seems to be taking forever to slew to its target, check the batteries. These are all classic symptoms of low power. Regardless, you'll have to run the setup sequence once again.

Speaking of batteries — what happens to a Go To when it's powerless? The more sophisticated scopes have a manual mode that lets you disengage the clutch system that normally drives the scope. Then you can use it as a regular telescope, moving it by hand and using its finder to locate objects the old-fashioned way — by looking for them! Some inexpensive Go To's simply seize up and are unusable without power, so make sure you always have spare batteries on hand. Not surprisingly, batteries and slewing motors don't like freezing winter weather — one more reason to choose a Go To scope with manual capability.

DON'T BE SHOCKED

Although most Go To's and other powered telescopes run off batteries, there are times when you might need additional electricity — for a computer, some auxiliary gear, or the scope itself. If this involves any type of insulated cord or flexible wire, check it often for fraying, breakage, or loose connectors. Immediately replace or repair any defective part or cord to prevent shock hazards or damage to your scope. Remember, electricity and water don't mix, so keep your power supplies dry and dew-free.

An Equatorial Balancing Act

Earlier I mentioned that a German equatorial mount (GEM) isn't the best one to have if you're just starting out. But if that's what you've got, here's something to keep in mind: a balanced telescope is a happy telescope.

To reduce the wear and tear on internal gears (and the motor drive if it has one), and to help the scope travel smoothly as it either tracks an object or moves from one target to another, make sure the scope is balanced. It's likely the manual tells you how to load the counterweight and balance the scope, but I feel it's worth mentioning here because a balanced GEM is a thing of beauty.

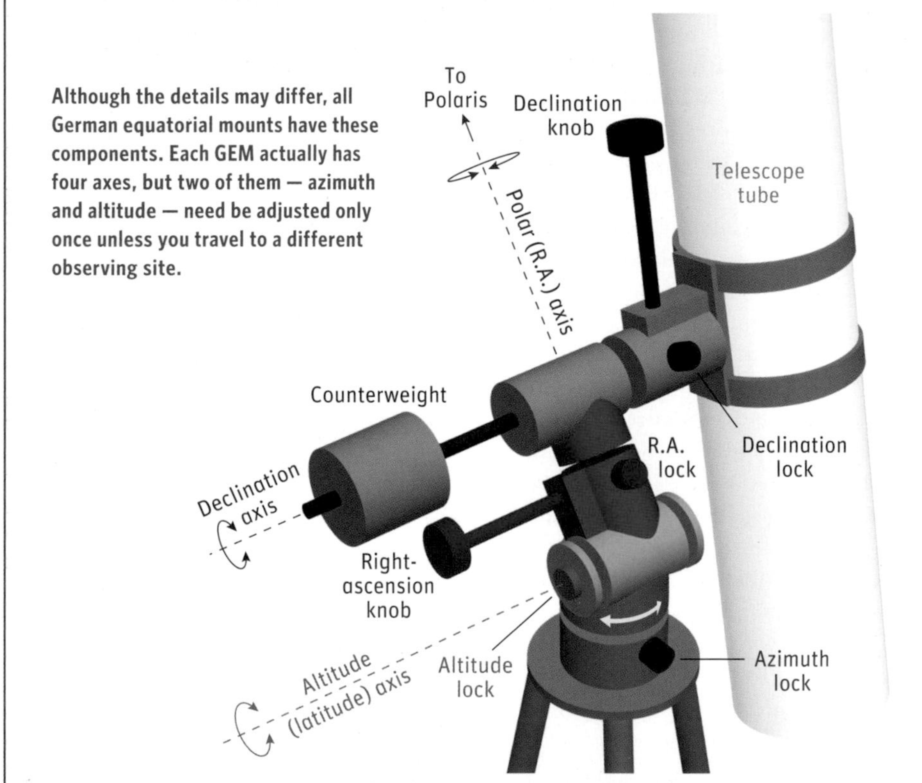

Although the details may differ, all German equatorial mounts have these components. Each GEM actually has four axes, but two of them — azimuth and altitude — need be adjusted only once unless you travel to a different observing site.

FAST ASSEMBLY

When putting your telescope and GEM away for the night, you'll find that they're easier to move if you break them into three major parts — the tube, the counterweight, and the mount itself. So for quicker assembly the next time out, especially if you always use the same setup and accessories on your scope, use a permanent marker to make a small mark on the tube to indicate where the tube rings rest and on the declination shaft to show the counterweight's position.

With the counterweight in place on the declination axis, loosen the right ascension (RA) clamp. You'll find it on the main body of the mount, above the tripod head. Adjust the counterweight up or down so that when the scope is pushed to one side, it stays in place and doesn't keep going or swing back. When you've done that, tighten the RA clamp. By the way, never allow the counterweight to rise higher than the scope tube. If it slips on the shaft, it could damage the mount.

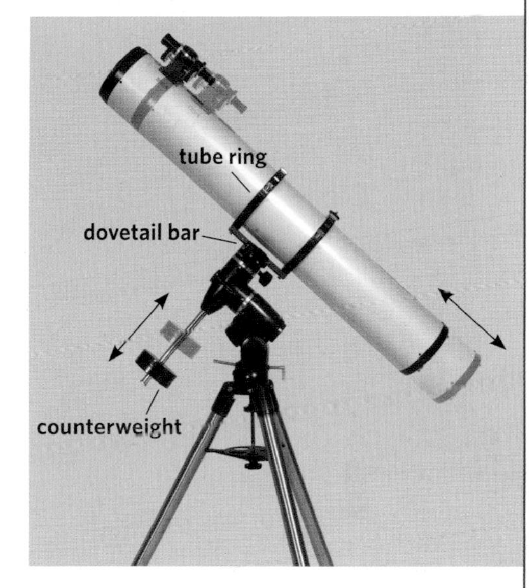

But for perfect balance, this act has a second part. Locate and loosen the declination clamp. You'll find it just under the tube on the same axis as the counterweight. Now loosen the screws on the tube rings just enough so that the tube can turn and slide. With an eyepiece, finderscope, and any other tube-mounted accessories in place, slide the scope back and forth in its mounting rings until the tube doesn't swing in declination. When the scope stays where you leave it, tighten the tube rings and then the declination screw.

POLAR ALIGNMENT

If a mount is polar aligned, it means that the mount's polar (right ascension) axis is pointed directly at Polaris, the North or Pole Star. (In reality, the polar axis should be aimed directly at the north celestial pole, but Polaris is very close to that spot.) If you use either a motor or a slow-motion control to gradually swing the scope around the polar axis, the sky's rotation is neutralized and an

For a quick initial alignment that's perfectly acceptable for visual observing, point the mount's polar axis toward Polaris, aim the scope parallel to the polar axis, and use the view of Polaris through the finderscope to fine-tune the altitude and azimuth adjustments so the North Star eventually shows up in your lowest-power eyepiece.

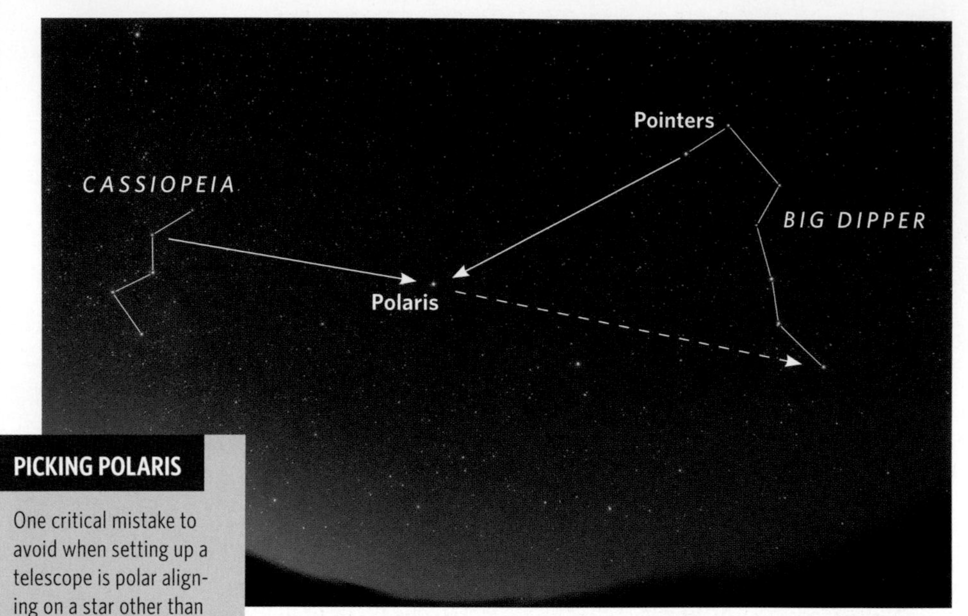

The Northern Hemisphere is blessed with a bright star — Polaris — that marks the location of Earth's north celestial pole. Use the Big Dipper or the W of Cassiopeia to find it.

PICKING POLARIS

One critical mistake to avoid when setting up a telescope is polar aligning on a star other than Polaris. Take your time and be certain you've got the right star. It sounds ridiculous, but mistakes happen. It's easy to confuse some of the other stars in the Little Dipper with Polaris, especially if tall trees or clouds are hiding portions of the northern sky.

object is tracked so it appears motionless in the eyepiece (even as the Earth turns under your feet).

Here are a few things to keep in mind that'll help smooth the alignment process. Find a flat, level spot for your scope, and try to pick a place where trees or houses don't hide Polaris. Select an area with enough walk-around room that you won't accidentally bump your scope and knock it out of alignment. Want a fast way to align your GEM? The image on page 21 shows you how. Finally, if you can't see Polaris, use a compass to find north. But beware. Magnetic north can differ significantly from true north (Polaris north). Still, if you're trying to observe a special celestial event and you can't see the North Star because of clouds or trees, a scope aligned to magnetic north will be better able to follow the sky than a nonaligned one. If you're doing only visual observing, an approximate alignment should be satisfactory.

Improve Your View

Collimation. There's a word that strikes fear into the heart of many new telescope owners. It shouldn't. Collimation just means aligning the center of your telescope's primary mirror or lens so that it aims at the center of the scope's eyepiece. Why bother? Because a poorly collimated instrument gives blurry views of everything and usually disappoints the user. You want the best views possible,

A properly collimated telescope will show stars as sharp points of light ***(right)***, not as fuzzy, offset glows ***(left)***.

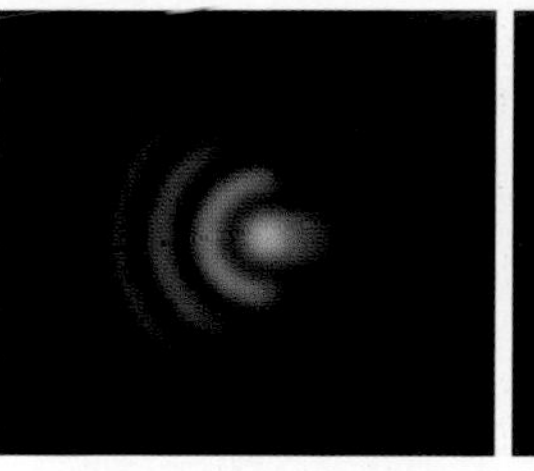

and that goal shouldn't be thwarted by something that's easy to fix. In fact, I think this is so important that I'm going to temporarily abandon my "This book is not a manual" mantra and go through the collimation process step by step — it *does* make a difference!

Every type of telescope needs to be collimated to give the best results, but a refractor, with its single objective lens at one end of the tube, seldom needs any adjustment. On the other hand, a reflector has two mirrors inside its tube that must be perfectly aligned in order to perform properly. The secondary, at the top (open) end of the tube, usually isn't a problem. But the primary mirror is mounted in a cell that holds it securely at the rear of the tube, and everything from unpacking your new scope to vibrations as you drive to an observing site can cause the mirror to shift out of alignment. So if your telescope has a life under the stars and is used often, it's going to get bumped around and knocked out of collimation now and then. But never fear. You don't need either a PhD in optical mechanics or a garage full of special gadgets to collimate your scope. Just a few simple tools and a little bit of patience are all it takes to align your telescope and make it perform like a champ.

Every reflector should come with one of these little collimation caps (available from Orion Telescopes & Binoculars). Still, making one from a 35-mm film canister isn't difficult.

MAKING A COLLIMATION TOOL

Before you begin, you'll need a collimating tool. Some scopes come with one or you can always buy it, but a simple tool can be easily made. Can you slide an empty 35-mm film canister into your eyepiece holder? If so, then drill a small 1⁄16-inch hole in the center of the container's cap. Next, use a razor or scissors to carefully trim off the bottom of the canister. The result is a peephole that will slip nicely into the telescope's focuser. Your newly made "collimation tool" is ready to use.

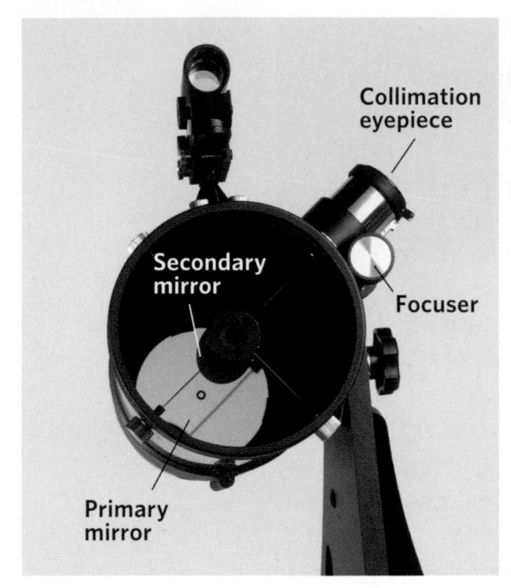

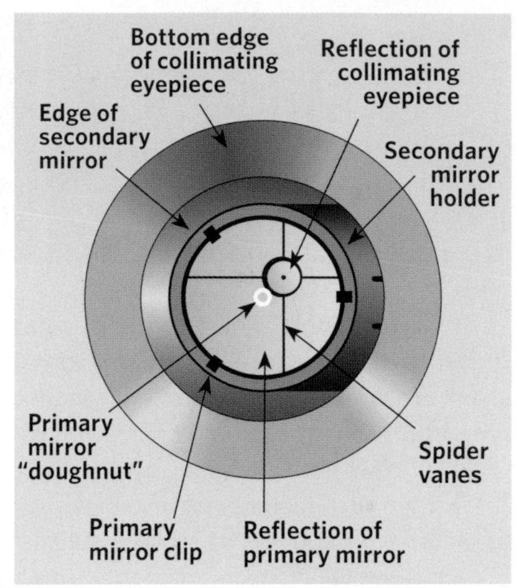

Before starting, make sure you identify everything you see when looking through the collimation tool *(right)*.

FIRST THINGS FIRST

Now is not the time to be multi-tasking. Concentrate on each step of the process and don't worry about the lawn or the laundry. And collimate during daylight — it's easier when you can see what you're doing.

With the collimating tool in the focuser where an eyepiece would normally be, point your telescope tube at the sky. Rack the focuser in as far as it will go and then look into the peephole. You'll see a lot of confusing reflections. Take your time identifying everything — the two important reflections are those of the collimating eyepiece visible in the primary mirror and the primary mirror's center mark.

Many mirrors have a center dot or doughnut marked on them. Your goal is to place the little peephole you're looking through right in the middle of the primary mirror. Are the peephole and dot centered? If so, don't do anything — your telescope is collimated. If not, then it's time to become familiar with the back end of your scope and perhaps enlist the aid of a helper.

Look at the back of your telescope, which is also the back of the mirror cell. You'll see a number of knurled knobs or screws — possibly three, but more likely six *(left)*. Scopes with six use three for mirror adjustment and three to keep the mirror in place. Check the scope's manual for clarification on which ones to adjust; otherwise you may be turning a knob that releases your mirror!

COLLIMATION 101

Start the collimation process by slowly turning one of the knobs about a quarter turn and then looking through the collimation tool's peephole — or get your helper to turn the knob while you watch what happens. Does the peephole move closer to, or farther away from, the center dot? If it moves closer, turn the knob again in the same direction. If it moves farther away, turn the knob in the opposite direction.

After you've turned the first knob and are as close to centering the peephole in the center dot of the primary mirror as possible, pick a second knob and repeat the process. Slowly move the second knob back and forth to get the peephole as close as possible to the center of the primary. If you can't quite get the peephole and the dot to align, return to the first knob and tweak. At this point you're fine-tuning the alignment, so take your time.

If your scope is badly misaligned, you may need to start with one knob to get the mirror back into the ballpark, and then use the other two to finish the alignment. But in most cases an almost-perfect collimation can be done with just two of the knobs. Be patient, and remember to make only slight turns on the knobs at each step.

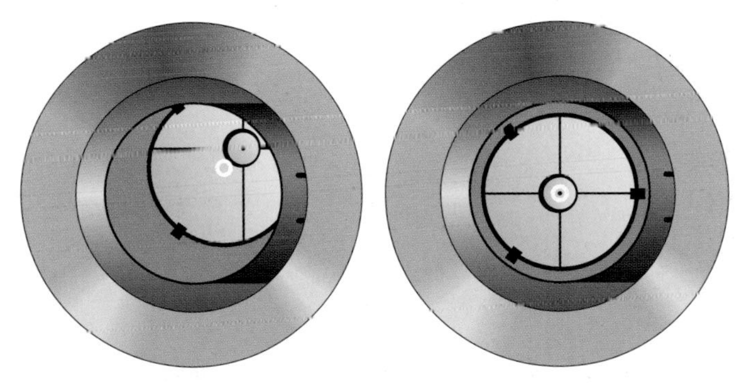

Your goal is to get the view through the collimation tool to go from a misaligned mirror *(left)* to one that's collimated *(right)*.

NO CENTER DOT?

What if there's no dot or ring at the center of your primary mirror? If the telescope is badly misaligned, you can probably eyeball it back into some semblance of collimation without a central doughnut (see the image above right and imagine it without the center dot). Then, to perfect your collimation, enlist the aid of a helper. Take your scope outside and find a bright star using low power. Switch to high power, center the star, and put it out of focus. Next, with your helper slowly moving the collimation knobs one at a time, try to get the star so it looks like the right-hand image on page 23. Be patient; the improved views will be worth it.

A FINAL TEST

Once collimation is complete and the stars begin to come out, it's time to try a star test. Use low power to find a bright star, then switch to a higher power and center the star in the field of view. If your collimation is perfect, the star in your telescope should be a tiny dot of light. If the collimation is off, the focused star will look smeared and not the nice neat dot it should be.

Now go past focus (in either direction) and see if the out-of-focus image of the star is surrounded by a series of symmetrical concentric rings. Think of the rings of a bull's-eye. When you see symmetrical rings inside and outside of focus, you're right on target. If they're offset (see the image on page 23), then your collimation isn't dead on. You can either take the scope back inside and collimate again or enlist someone's help and follow the instructions listed in the section "No Center Dot?"

After you've collimated your optics a few times, it's not scary at all and is definitely worth the extra few moments it takes to get the best views your telescope has to offer.

Telescope Tune-up

At the start of this chapter I suggested that if you have an old telescope in the back of your closet, take it out and give it another try. If you do, you may quickly rediscover why it was relegated to the closet in the first place—it just doesn't work very well. Is it wobbly and hard to aim? Does even the Moon look disappointing through it? Let me suggest a few simple ways to turn this dust collector into a working telescope.

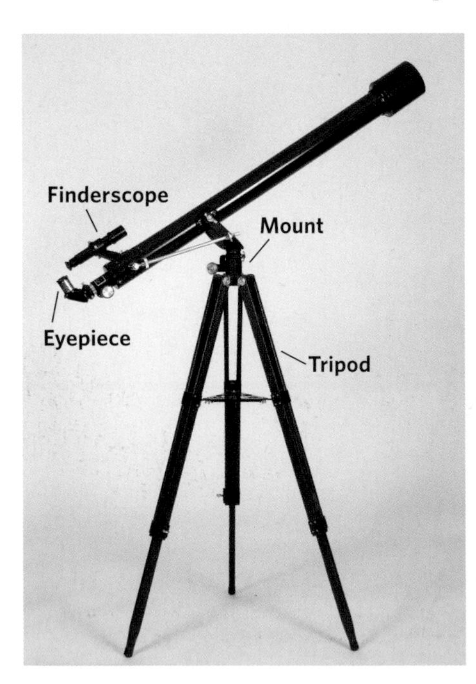

Often the worst feature of an inexpensive little scope is its wobbly mount. Check all the fasteners holding the mount and tripod legs together, and tighten everything you can. Replace the nuts and washers on the tripod with the locking variety for added holding power. If it still seems shaky and wobbles with even the slightest touch, maybe the unit's center of gravity is too high. Fill a plastic bag with rocks, or an old gallon jug with sand, and suspend it by a cord from the base of the mount through the tripod's legs. For more stability, consider not extending the tripod's legs to their maximum and observe sitting instead of standing. Finally, make sure that the nuts holding the scope to the mount itself are tight.

If your mount has slow-motion adjustment cables that are long and springy, swap them for shorter, thicker cables.

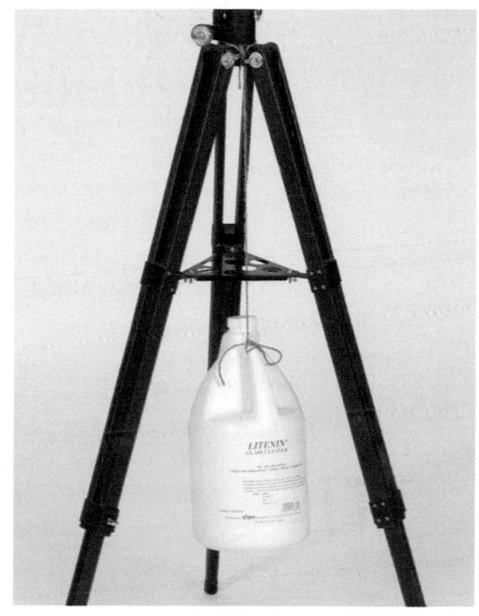

An easy way to improve a tripod's stability is to hang a weighted jug between its legs.

Another problem area is likely the finderscope — the little mini-telescope that attaches to the side of the main scope and helps you find things to look at. See page 37 to discover how you can improve (or replace) it.

Finally, if the scope is equipped with tiny 0.965-inch eyepieces, get rid of them. Buy an adapter that will allow the telescope to use the more common 1¼-inch eyepieces. This alone should significantly improve the image quality, since most of the 0.965-inch eyepieces are of extremely poor quality. Keep in mind that you'll not be using the full diameter of the 1¼-inch eyepiece, but at least you will be looking through better optics. And stick to low-power eyepieces. High power tends to bring out all the defects in a telescope. Besides, with moderate magnification you won't need to chase the object across the field of view as often, so you can concentrate on observing rather than tracking. Turn the page for a full discussion of eyepieces.

If you're using an adapter tube *(left)* for 1¼-inch eyepieces in a telescope built for 0.965-inch eyepieces, make sure the scope's focuser can be racked in far enough to focus the image *(right)*. A more expensive option is to buy a special diagonal that accepts 1¼-inch eyepieces but can be inserted into your telescope's 0.965-inch focusing tube.

Exploring Eyepieces and Filters

Ask three of your astronomy buddies what their favorite eyepiece is, and you could end up in the middle of a lively discussion. That's because there are so many eyepiece choices available that it's unlikely your three friends have even one in common! It would certainly make things easier if I could just say, "Use a 32-mm wide-field and you're all set" (that's *my* favorite, in case you couldn't guess), but that's not how it is.

Eyepieces come in three sizes. The small 0.965-inch ones are no longer common (see page 27 if you have them), while the large 2-inch eyepiece is becoming more popular. But the 1.25-inch (middle) is the worldwide default standard.

EYEPIECE TYPES

Eyepieces come in three different sizes and many different designs. When you hear names like Huygenian, Ramsden, Kellner, Orthoscopic, Erfle, Plössl, Nagler, and Panoptic being tossed about, you know that eyepieces are being discussed. Some are cheap — Huygenian and Ramsden usually come with inexpensive starter scopes (look for the *H* or *R* on the eyepiece) and should be replaced. Others are expensive — paying $300 for one eyepiece isn't unusual. Generally, you get what you pay for. Why? Because the more elements (or lenses) in an eyepiece the better it is and the more expensive it is to make.

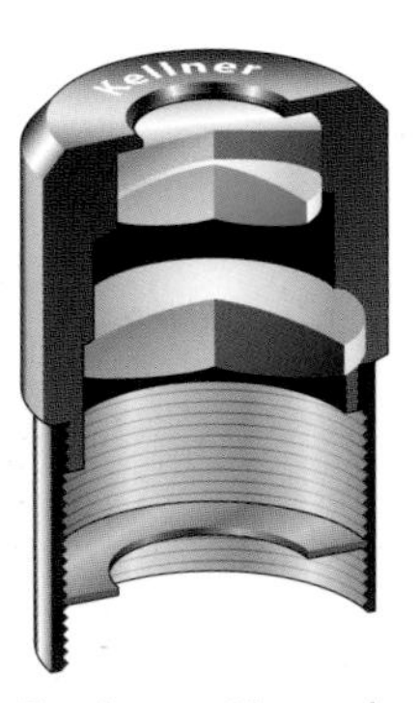

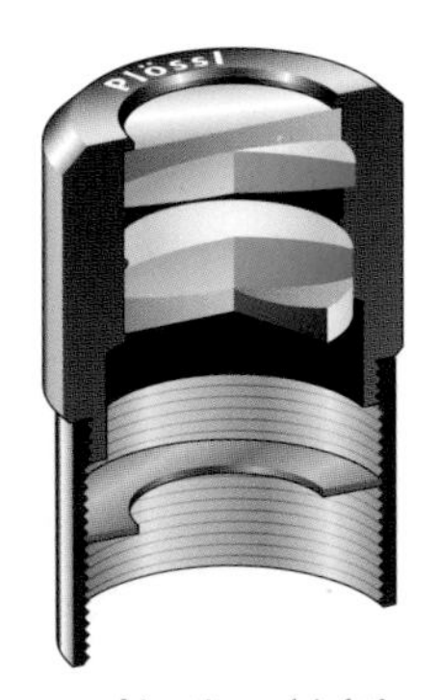

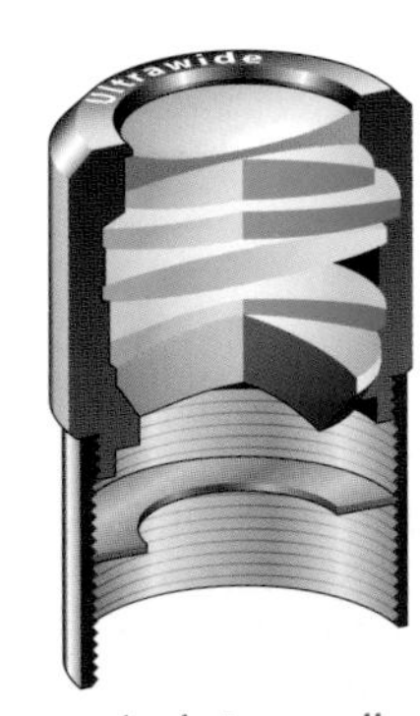

Eyepieces with complex lens combinations *(right)* are expensive but generally offer superior views. Some eyepiece designs work better with different types of scopes. Simple eyepiece designs — those with four elements or fewer, like Kellner *(left)*, Plössl *(middle)*, and Orthoscopic, for example — don't do well on short-focal-length telescopes.

TOO MUCH POWER?

On the outside of boxes of "El Cheapo" telescopes you'll often find the phrase "Magnifies 500 Times!" Technically, you *can* get 500× from a 60-mm (2.4-inch) refractor — you just won't see anything because your target will look like a huge blur. There's a handy rule of thumb for determining the maximum power for your telescope: no more than 50 times the aperture of your scope's main lens or mirror if measured in inches, 2 times the aperture if measured in millimeters. For that 60-mm scope, this means no more than 120×. Even so, you're not going to be able to use this much magnification unless conditions are perfect — you'll probably have best results using half that power.

Buy the best-quality eyepieces you can afford. As the years pass you'll likely change telescopes, but first-class eyepieces will last a lifetime and can be used in almost every scope. If you're thinking of buying several different eyepieces, consider acquiring a complete set that's *parfocal*. That means each eyepiece comes to a focus at almost the same point, requiring little or no refocusing when you switch eyepieces. You can't create your own parfocal set by combining different types or brands; you'll have to buy the eyepieces as a complete set from a particular manufacturer. (Instead of numerous eyepieces, consider a zoom lens; turn the page to learn more.)

When you're at the telescope, make sure the setscrew that holds the eyepiece secure in the focusing tube (or star diagonal) is tight. You don't want the eyepiece to drop out! Some eyepieces have fold-down eyecups that help block stray light — use them. (You can buy these eyecups from most telescope retailers.) And notice that almost all eyepieces have threads (on the end opposite the lens) for adding filters, which opens up another world of viewing options, not to mention outright fun. Turn to page 31 for more details.

QUICK FOCUS

If you'd like to explore the Moon during its thin crescent phases in bright twilight or find Venus (or even the Moon) in broad daylight, you'll quickly discover that achieving a good focus can be difficult. So one night when the seeing is steady and the focus is perfect, use a felt pen to make a thin mark on your scope's focusing tube where a couple of your favorite eyepieces come to focus. Now, no matter what you're after — or when — just insert your favorite eyepiece, rack the focusing tube to your mark, and you're ready to hunt.

ZOOMS AND BARLOWS

A generation ago zoom eyepieces were considered novel but poor-quality accessories. No longer. High-quality zooms, which change magnification with a simple twist of the barrel, have become very popular. Some have click stops at certain focal lengths so you can easily determine what power you're using. You can zoom out for a wide-field view to hunt an object, and then zoom in for a closer look once you've found it. When thought of as a single eyepiece, a quality zoom is expensive. But in reality, it can replace a set of three or four eyepieces, making a zoom a great bargain.

If you don't own an assortment of eyepieces and a zoom is too expensive, try a Barlow lens to augment your viewing options. When placed between the telescope and the eyepiece,

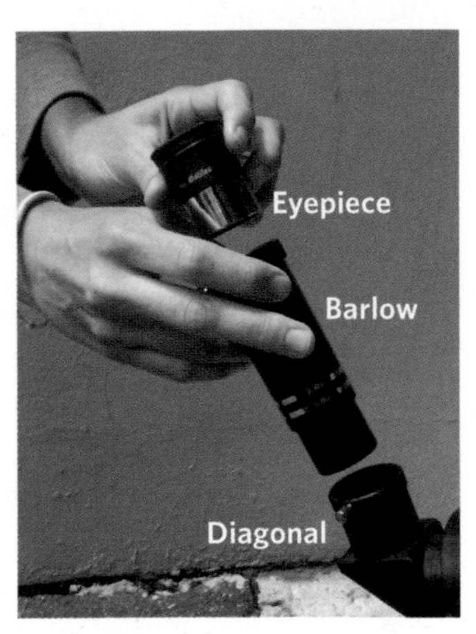

Ask veteran observers what their most valuable accessory is, and you'll frequently hear, "a Barlow lens," because it lets each eyepiece in their collection do double duty.

A zoom *(right)* can take the place of several regular eyepieces and is perfect for use with a travel telescope. Note the focal-length indicators at the click stops on this particular zoom.

WHAT'S MY MAGNIFICATION?

A question I'm constantly asked by new observers is: "How do I figure out the magnification of my telescope?" Actually, that's the wrong question. It should be: "What's the magnification of this eyepiece when I use it with my telescope?" You change magnification by swapping one eyepiece for another.

To find the magnification of any particular eyepiece, you need to know the *focal length of the eyepiece* (it's written on the barrel and is always in millimeters) and the *focal length of the telescope* (usually found somewhere on the tube, on the focuser, or in the manual). If the telescope's focal length is in inches, multiply the inches by 25.4 to get millimeters.

Divide the focal length of the telescope (a big number) by the focal length of the eyepiece (a small number). The result is the magnification for that particular eyepiece. If your little reflector has a 1,200-mm focal length and you have two eyepieces marked 25 mm and 10 mm you'll get magnifications of 48× and 120×, respectively (the "×" stands for "power"). That's it! If you use these same eyepieces on a different scope with a focal length of 2,000 mm, then you'll get magnifications of 80× and 200×.

the Barlow increases the magnification of the eyepiece. A 2× Barlow doubles the magnification of an eyepiece; a 3× triples it. If you're just starting your eyepiece collection, purchase a good-quality Barlow, and then add eyepieces in such a way that when you apply the Barlow to each eyepiece, it gives a new magnification, not an existing one. For example, there's no point in having 24-mm and 12-mm eyepieces since applying a 2× Barlow to the 24-mm produces the equivalent magnification of the 12-mm eyepiece. Better to have a 24-mm and a 10-mm instead. Regardless, Barlows aren't expensive, so don't scrimp. There's no point in using a cheap Barlow with an expensive eyepiece.

EYEPIECES AND EYEGLASSES

To wear or not to wear — eyeglasses at the eyepiece, that is. If you're near- or far-sighted, the answer is no: just take them off and refocus the telescope. But if you suffer from astigmatism, then you'll need your glasses for sharp views.

FILTERING YOUR VIEW

Filters, like eyepieces, are guaranteed to stir passionate discussion among veteran observers. They seem to come in every color of the rainbow, and no single filter will work best for every object. See the image on page 32 for an example of what two different colors can do. A filter that's often included with small telescopes is a Moon filter, which reduces light coming from the Moon without changing the color. (A better filter for the Moon is a variable polarizer.)

In addition to color filters, there are three other types of filters, all of which are useful for deep-sky observing because they block skyglow and increase the contrast between the nebula you're trying to see and the background sky. But as you might guess, they're more expensive than the simple color filters.

Broadband filters are also called light-pollution reduction (LPR) or deep-sky filters because they block the glow from mercury va-

Most of the filters you'll encounter are threaded and screw into the base of an eyepiece.

Here's a typical starter set of four color filters. The filters have numbers — #38A, #58, and so on — that indicate different colors. These are Wratten numbers, named after a system developed by Kodak in the early 1900s.

Colored filters make it easier to see details on objects like Jupiter. At left is a non-filtered view of the giant planet; the colors are quite subtle. The middle image shows a simulated view through a red filter; at right is a blue-filter view. Note that some features are more prominent when seen through one filter or another.

por, high- and low-pressure sodium vapor, and neon lights. They also reduce skyglow, though they won't eliminate it. A broadband is a great all-around LPR filter for citybound stargazers and may be just enough to let you coax a nebulous object out of hiding in a light-polluted sky.

Narrowband filters are designed for looking at nebulas glowing in green and bluish green light. They work well under light-polluted skies. You'll also hear them referred to as UHC — ultra-high-contrast — filters.

Emission-line filters are very specialized — think of them as narrow narrowband filters. An example is an oxygen III (or O-III) filter, which makes some nebulas much easier to see. But it's so specialized that it can actually dim other nebulas.

Broadband, narrowband, and emission-line filters are no substitute for a dark country sky, but they can improve your view of

SAFE SOLAR VIEWING

There's another filter worth mentioning: a solar filter. Safe solar filters are designed to fit securely over the *front* of the telescope, where the light first enters the instrument. They're available in many different sizes; you can even get a pair of small filters to fit over the lenses of binoculars.

Never use a solar filter that screws into an eyepiece. Some older, inexpensive telescopes used to be sold with such a filter. If you have one, **throw it away!** It can crack unexpectedly and your eye could be damaged due to exposure to magnified unfiltered sunlight.

One more thing. If you're using your properly filtered telescope to observe the Sun, cover your finderscope. You don't want to accidentally look at the Sun through even this mini-telescope.

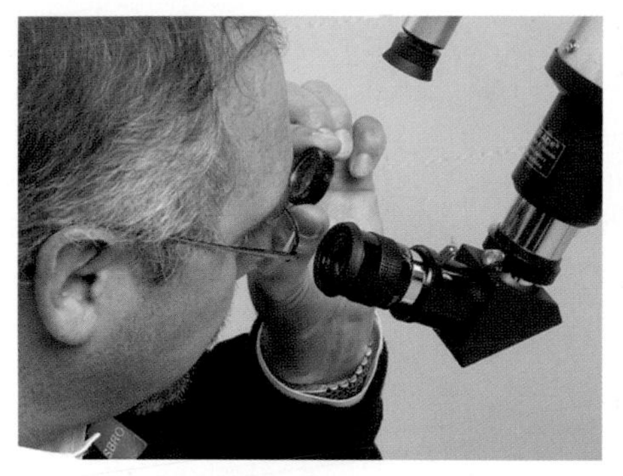
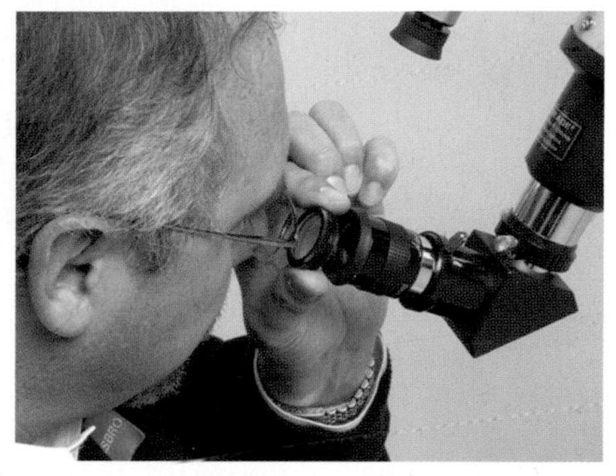

If you're trying to spot a faint object using a broadband filter, try holding the filter and flicking it back and forth in front of the eyepiece. Alternating between the filter in place and not in place may cause the object to pop into view. This trick is also useful when you're deciding which colored filter to use; hold it in front of the eyepiece to see what the view is like before taking the time to pull the eyepiece out, screw in the filter, and reinsert the eyepiece into the focuser or star diagonal.

many objects if you're stuck in the city. Even when used at a rural observing site, they're helpful for boosting contrast and reducing glare. In a city sky, they can sometimes make the difference between seeing or not seeing a faint nebula. If you've joined an astronomy club (see Chapter 4), perhaps you can borrow different filters to experiment with in different situations to determine which ones work best at your observing site.

Keep It Clean

According to an old adage, an ounce of prevention is worth a pound of cure — a little precaution now is preferable to a lot of cleaning up later. If you're a new telescope owner, now is the best time to get a clean start. Good equipment-keeping habits today will save you grief, and possibly money, tomorrow.

When you store your telescope outside, whether permanently or just for a few days as here at the 2005 Texas Star Party, protecting your optics is a must.

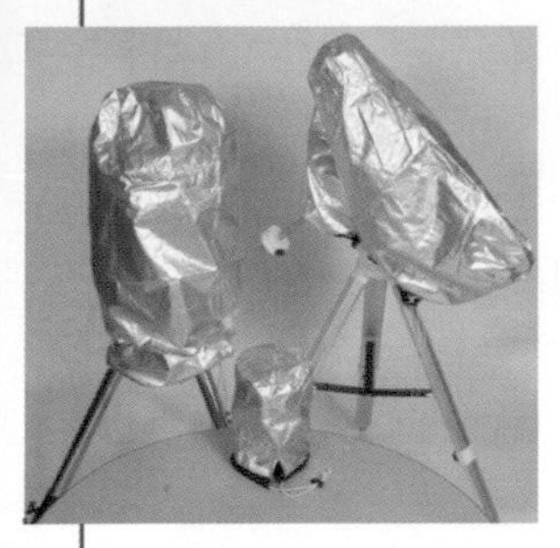

Though sometimes designed for indoor protection, not long-term outdoor use, telescope covers are great for protecting your scope if you store it inside for most of the year.

Always protect your telescope, even if it's stored indoors. A telescope cover can be purchased or made, and almost anything devised to help keep dirt, dust, and insects out of your scope will extend its lifetime. Even a shower cap placed over each end of a reflector's tube will help! Simple tarps with bungee cords, or patio-furniture covers converted to scope covers will work in a pinch. Don't forget to plug the eyepiece holder—a 35-mm film canister fits nicely into a standard 1¼-inch focuser. If you leave your reflecting telescope assembled, store it with the tube aimed very slightly down. This helps reduce the amount of dust that might settle on the mirror, and it ensures that nothing will accidentally fall down the tube and hit the primary.

Keep all your other astronomy gear covered or in cases. Invest in eyepiece containers or a foam-lined box to keep your optics safe. A small plastic utility, tackle, or toolbox can be purchased for next to nothing. Not only does this help protect everything, it ensures that all your stuff is together in one place.

EXCUSE THE DUST

Let's be realistic — a little dirt doesn't hurt. A layer of fine dust on the primary mirror of a reflector or on the front corrector plate of a compound scope isn't what most amateurs would consider "dirt." And while the dust layer may slightly affect your view, the image's quality is more dependent on such things as the sky conditions

An optics-washing crew uses natural sponges and soap to clean the 8-meter (315-inch) mirror of the Gemini North telescope in Hawaii. Such extreme measures aren't needed to clean amateur optics, but washing your mirror isn't a project to be undertaken lightly.

and the object you're looking at. Generally, if a mirror or corrector plate has a little dust on it, leave it alone. Follow the physician's general philosophy: First, do no harm. If necessary, a dirty mirror can be inexpensively cleaned; a scratched mirror due to a botched cleaning job costs far more to make right.

The bottom line is that cleaning a coated optical surface is potentially the worst action it will suffer other than breakage. *Sky & Telescope*'s Alan MacRobert has an insightful response to questions he often receives from beginners who want to know when a scope's optics need cleaning. "If you're asking the question, they don't." And if you're really concerned about the condition of your mirror, seek advice from an experienced telescope user.

CARING FOR EYEPIECES

By protecting your scope's mirrors (or the main lens if it's a refractor), you can observe for years without worrying about a major cleaning. But eyepieces are on the frontline when you're using your scope. Eyepieces suffer the most wear and tear from normal use — they're handled often, occasionally dropped, left in lint-lined pockets, and attacked by our eyelashes as we view through them. And if you become involved in public star parties (observing sessions), someone will inevitably leave a fingerprint on the eyepiece's lens after touching it while saying, "You mean look right here?"

To remove dust from an eyepiece, suck air between a finger and the lens. Don't blow on an eyepiece, since droplets of spittle can land on the lens and cause spots.

If your eyepiece looks a little dusty, brush it lightly with a camel's-hair brush (available in camera stores). If that doesn't work, use a can of compressed air to blow on the glass. Be careful not to shake the can or turn it upside down. You don't want the liquid propellant spitting onto the eyepiece, because you may wind up with more gunk on your optics than you started out with.

If an eyepiece has more stubborn goop and grime on it, then a thorough cleaning is in order. Lightly moisten but don't saturate a soft wipe or cotton bud with lens cleaner (both also available at photo-supply shops) and gently swirl it across the lens. Don't drop the liquid directly onto the glass and never use pressure since this may scratch the lens. Repeat the process a second time; then quit.

By the way, this eyepiece-cleaning process can also be applied to the eyepieces and lenses in binoculars.

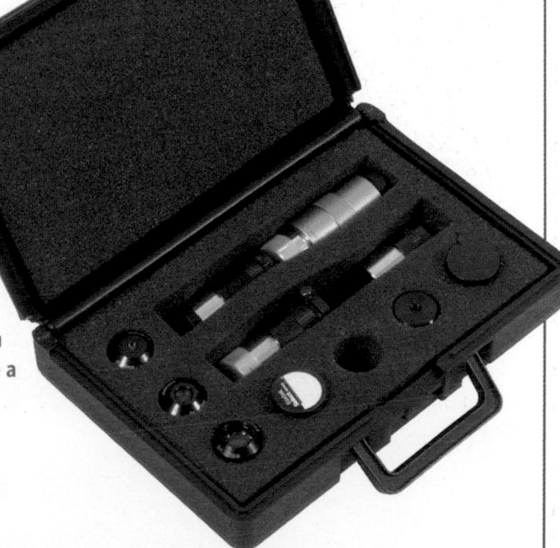

No matter how many eyepieces you own, keep them safe (and clean) in their own case. If you don't have a lot, use the extra space to store other accessories.

Finding Your Way

Most telescopes come with finderscopes (finders)—mini-telescopes mounted on the side of the main tube that show a wider swath of sky than does the big instrument. Finders come in a variety of sizes and shapes and have different levels of magnification. They're wonderfully useful, and yet one of my pet peeves is the inadequate finder that often accompanies commercially made telescopes. Budding amateur astronomers don't have skeletons in their closets; they have abandoned telescopes with terrible finders!

The worst offenders are the 5 × 24 finderscopes that often come with small scopes. (The first number is the finder's magnification and the second number is its aperture in millimeters.) These usually have a plastic lens, bad focus, and a poor mounting bracket. If your telescope has one of these, discard it. Seriously.

But not all finders are evil contraptions. Many scopes now come with a one-power (1×) finder — often called a *reflex* or *unit-power finder.* (It's so named because it doesn't magnify anything.) These one-power wonders are very intuitive to use, since the viewing window is transparent and doesn't alter the view of the sky shown behind it. Instead, an illuminated red dot or a red bull's eye is projected onto the window (and therefore seemingly onto the sky) to help center the target in the scope's eyepiece.

If you have a small telescope, a Go To scope, or a large instrument camped under dark skies, a unit-power finder is a fine choice. But if your sky is badly light polluted, a finder that magnifies might be a better option. These come in many varieties, including 6 × 30, 7 × 50, 8 × 40, 9 × 50, and even 10 × 50. Of course, just because a finder isn't one of the dreaded 5 × 24s doesn't mean it's perfect; you might have to perform some creative upgrades on any of these finders.

WIDE FIELD, NARROW FIELD

If you can't find it, you can't observe it. It's almost impossible to locate anything using just the telescope itself (except the Moon and some bright stars or planets) because, even with a low-power eyepiece, your main scope shows only a tiny part of the sky. A finderscope has a much wider field of view, making it easier to locate celestial sights. The 1° circle around the Pleiades shows a typical field of view of a low-power eyepiece in a telescope; the 5° circle is the minimum amount of sky that you should see in a 6 × 30 or 7 × 50 finderscope.

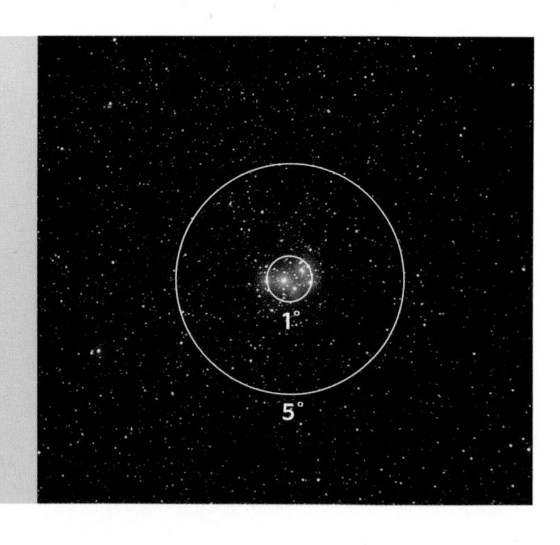

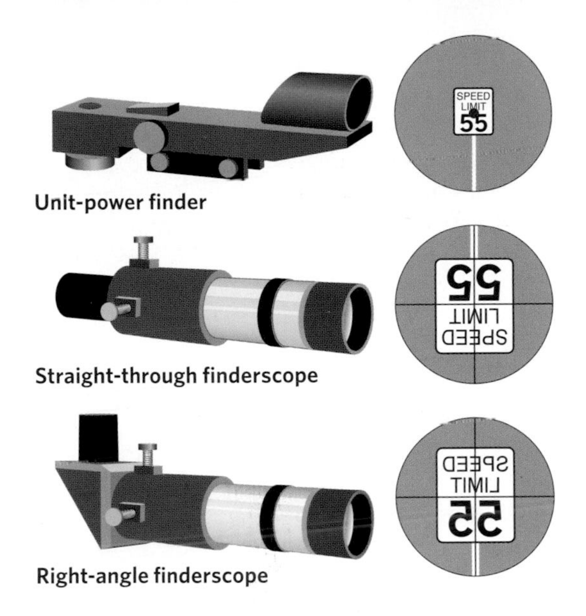

First-time telescope users often comment: "What I see in my finder is not what I see in my scope." Assuming the finder and telescope are aligned, the difference lies in the optics that both instruments use (see page 12 for the view through different telescopes). Unit-power finders *(top)* show a "normal" sky — the one you see without optical aid. Most straight-through finders merely turn the image upside down *(middle),* while right-angle finderscopes *(bottom)* mirror-reverse the upright image.

FINDER ALIGNMENT

If you need to realign your finderscope (which you should also do after collimation), do it during daylight so you're not fumbling in the dark with adjustments or hunting for dropped thumbscrews.

Select an object at least several hundred yards (or meters) away — like a tower or power pole. Sight down the telescope tube and, using your lowest-power eyepiece, get the object in the field of view of the main scope. Now insert your highest-power eyepiece and center the object, then lock your scope so it doesn't move. Next, look through the finder and determine how far away the target is from the center crosshairs. One by one, adjust the screws

---

MAKING A BAD FINDER BEHAVE

To make the process of aligning finder and telescope easier, consider replacing the tiny adjustment screws with thumbscrews or knurled screws so they can be turned easily by hand during alignment. If you observe during cold weather, large screws are easier to turn when you're wearing gloves.

If the screws that attach the finder's mount to the telescope tube keep coming loose, tighten them and then dot them lightly with clear nail polish. This will help hold them in place, though not permanently as would glue. Avoid putting any kind of adhesive on the screws that hold the finderscope in its mount, as these are always meant to be adjustable.

If the finderscope keeps slipping in its mount, wrap a little tape around the finder's tube — it will provide something for the adjustment screws to grip.

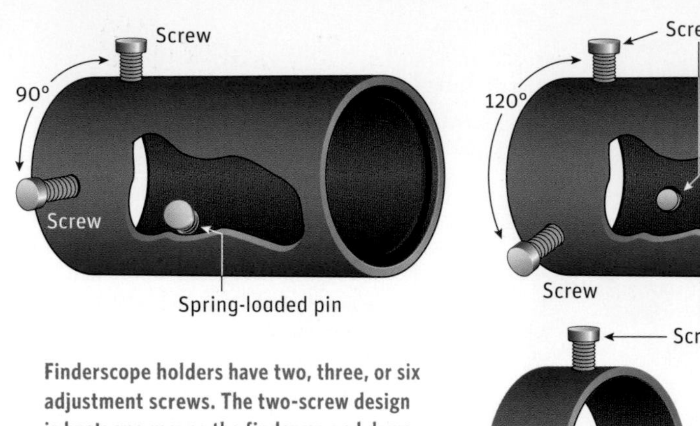

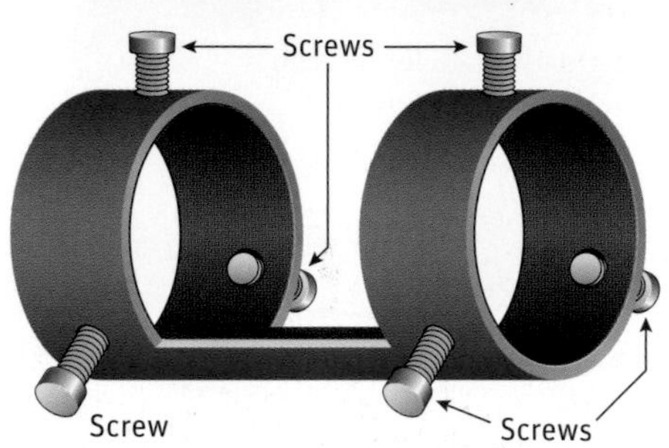

Finderscope holders have two, three, or six adjustment screws. The two-screw design is best: one moves the finder up and down, the other left and right. If yours has three or six screws, make sure they're all tightened (but not too tight!) before you begin the alignment process.

that hold the finder in its mount (not the screws holding the mount to the telescope). It's a lot like collimation — slowly turn one of the screws while looking through the finder to see what happens. It's a bit hit or miss, but eventually you should be able to center the target.

Now pick another distant target, this time using only the finderscope. If the finder is aligned, the object should appear in the eyepiece. If not, then adjust the telescope until it is, and tweak the finderscope's alignment until the target is centered.

If your scope has a unit-power finder, the process is the same. Instead of thumbscrews, these finders have small buttons on their sides or back that let you adjust them up and down or side to side.

A Scope to Go

Whether it's a trip to a distant corner of the world or to another locale in your own country, taking a portable telescope will keep you stargazing when you're away from home. Even if your journey is only to a nearby dark-sky site, factors such as a scope's size, weight, and need for battery or electrical power become important concerns. And don't forget setup time. After arriving at your destination, you'll want to minimize the amount of time spent fussing over assembly and adjustment, and maximize the time spent stargazing. Once you start contemplating telescopes for travel, you'll find that a surprising number of small scopes are actually travel-ready instruments masquerading as stay-at-home scopes.

The first consideration is your method of transportation. If you're going to a remote site by car, your travel scope can be any-

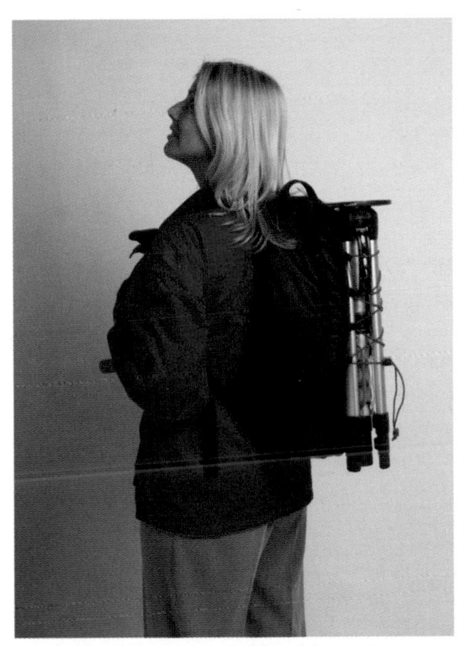

Some telescope manufacturers are now offering complete travel kits *(above, right)* that contain the scope, accessories (including a tripod), and a pack.

A customized travel case *(left)* or a regular hard-shell travel case available from any photo-supply retailer is the best way to protect your small scope if you want to take it traveling. Don't forget to leave room for accessories!

thing that'll break down and fit into your vehicle! If you're taking a driving vacation, pack the telescope first, not last, because it's not something you're likely to need right away.

But different rules apply when you're flying. If your scope goes as checked baggage, understand that it will not be handled with the same amount of care *you* lavish on it. After all, to a baggage handler it's just another piece of luggage. For this reason, I highly recommend that you take your precious telescope as a carry-on item. Now your choice of scope has a significant restriction: the

SPOTTING SCOPES

Another option for a grab-and-go instrument is a spotting scope — a compact refractor used primarily for terrestrial viewing. I'd never suggest having one of these as your main telescope, but they're compact, lightweight, and easily slipped into a small pack or bag with space left over for extra supplies. (Remember, you'll still need a tripod to mount it on.) Some spotting scopes come with a zoom eyepiece, which is very handy for travel even if it's not the best quality.

airline's carry-on baggage allowance. You'll also need to carefully cull your accessories down to the bare essentials, though a tripod and other non-optical components (including a chart, some tools, and extra batteries if required) can be checked. But don't go overboard; airlines also have weight restrictions on checked baggage, and you could end up paying an overweight fee.

MORE SKY WITH BOTH EYES

If all this sounds like too much hassle, don't despair. I have another suggestion for a portable skywatching instrument: binoculars. Think of them as a pair of joined mini-telescopes. Binoculars are lightweight and small enough to slip into a carry-on bag, backpack, or tote bag — with room to spare for a pocket star atlas and a red-light flashlight. Binos often come in a padded case that'll absorb any bumps and thumps encountered during the journey. Setup time is nonexistent, and stargazing can be done from anywhere you can see the stars.

As an added bonus, binoculars are a perfect daytime trip companion. Whether you're bird watching, people watching, checking out distant scenery, or investigating faraway landmarks, binos are a wonderful, multi-use travel accessory that provides instant gratification for any observer.

Of course any binoculars are better than none. Check your closet; if you have them, dig them out and turn them skyward. But if you can afford to, consider image-stabilized (IS) binoculars. IS binoculars use electronics and moving optics to steady the view, making them an excellent, albeit expensive choice for stargazing.

If you have trouble keeping your binoculars steady while observing, lean against a building or prop your elbows on a fence rail, picnic table, or car. The best solution may be to buy an adapter so you can attach them to a camera tripod *(right)*.

A lounge chair is the perfect accessory for binocular viewing.

With the push of a button, the image suddenly becomes stable — it's as if a tripod has magically appeared underneath the binoculars. Eliminating the dreaded bino-jiggle lets you observe celestial sights fainter than you might think possible. In fact, 10 × 30 IS binoculars often show as much as hand-held 7 × 50s. They do need batteries to operate, so remember to pack extras when you travel.

BINOCULAR BITS

Look at any binocular and you'll see a set of numbers printed near the eyepiece: 7 × 35, 8 × 40, 10 × 50, and so on. The first is the magnifying power of the instrument; the second is the diameter (in millimeters) of the objective lenses (the large ones in front).

If you're using binoculars for astronomy and you have a choice, go for the 10 × 50s. The bigger the objective, the more light it collects, but binos with lenses larger than 50 mm are heavy and hard to hold. Mounting them on a tripod will help steady the image, but now you're starting to add equipment, and I think this takes away from the ease of using them. And magnifications higher than 10× produce a small field of view, which can make it challenging to aim the binos at a specific celestial target.

You can use the magnification and aperture numbers to estimate the relative performance of different binoculars by simply multiplying the two numbers together. So 7 × 35 binoculars would be rated at 245, while 10 × 50s rate at 500 and, of the two, would provide better views. (My favorite is an 8 × 56, which scores 448.) This reinforces the idea that magnification and objective-lens size are closely related and it's not simply a matter of picking binoculars with either the biggest lenses or highest magnification. (The exception: this rating system doesn't apply to image-stabilized binos.)

Secrets of Stargazing

You wouldn't head out on an all-day fishing expedition without a little preparation, would you? First you'd look at a weather forecast, then check the road conditions, and finally ensure that your gear was all together and in working order before you left the house. A spur-of-the-moment trip is nice, but planning ahead usually increases your chances of catching a fish.

Fishing for celestial targets isn't much different. You need to plan. Too often new stargazers head inside early because "there's nothing to see." Folks, there's an entire universe out there full of things to see, and a little preparation will help you find them.

WHO LET THE STARS OUT?

A dark, star-filled rural sky looks very different from the light-polluted city sky you may be used to seeing from your home. I've noticed first-time attendees at the Texas Star Party walk around awestruck by all of the stars they can see. Many find it difficult to locate their accustomed landmarks in the sky because there are just so many stars. In particular, novices often have a hard time finding deep-sky objects within the gentle glow of our Milky Way galaxy *(left)*.

Plan Ahead

It sounds simple enough, but planning will help you get the most out of each stargazing opportunity. As with any other activity or hobby, being successful (and having a good time) depends a lot on being prepared. Develop a checklist of objects you want to see and a possible order in which to view them. After all, you don't want to miss something that's setting in the west because you spent too much time gazing at something else rising in the east.

WHAT'S UP?

If you're a novice stargazer, the logical question to ask is: "How do I find out what's going on in the sky?" You can't make a list of celestial goodies to observe if you don't have a clue about what's up!

The quickest way to determine which constellations are visible (and when) is to use a planisphere, or "star wheel." By spinning the star wheel to a desired date and time, you'll immediately see which constellations are visible and which ones will soon rise (or set). Sure, there are computer programs that do this, but a planisphere is quick, easy, and very portable.

If you're particularly interested in spotting pretty scenes involving the Moon, bright planets, and bright stars, pick up an astronomical calendar for the upcoming year. In addition to listing

Always keep looking up. You never know what pretty sights the sky will reveal.

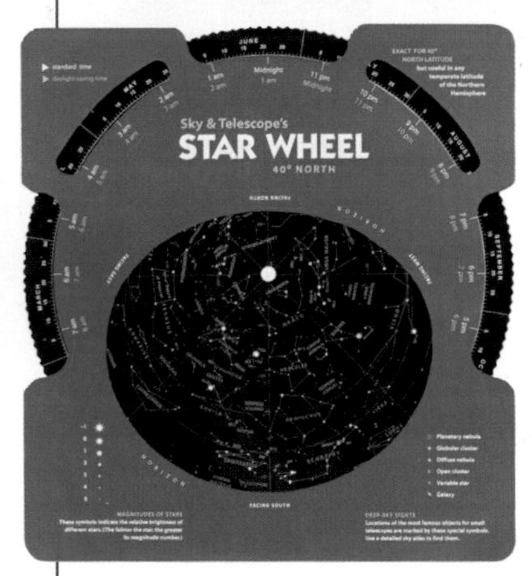

A planisphere will show you the heavens for any time and date; be sure to get one designed for your latitude.

these gatherings, a calendar such as *Celestial Wonders* will also tell you when to look for meteor showers, eclipses, and other predictable astronomical events.

Check to see if your local newspaper carries a weekly or monthly column about the night sky. Other sources of information include *Star Date* — a daily radio program — and *Star Gazer* with Jack Horkheimer, an internationally syndicated weekly television program that's also available via satellite and podcast.

For more detailed information about what you can see in the sky, it's still tough to beat magazines. *Sky & Telescope* is a monthly magazine written specifically for observers It has monthly star charts, observing tips, equipment reviews, and noteworthy articles (and images) highlighting the sky's best offerings. Other astronomy magazines published in Canada, the UK, Australia, and elsewhere contain similar fare.

Finally, there's always the Web. But beware — not everything you find there is accurate, up to date, or reliable. I've listed a few of the more reputable sites (plus some magazines and books) in Chapter 5 — the Resources section.

Of course, sometimes you *do* want to observe on the spur of the moment. I've been looking up for more years than I often care to admit, and occasionally I find myself without an observing plan. When this happens, I usually figure out what constellations are visible, turn to an appropriate page in my star atlas, and then try to hunt down every object shown on it. Since my favorite scope is not a Go To, I have to star-hop to find these sights (see page 63 for more about star-hopping), which usually takes up most of the night. It's fun and frequently leads me to something I've never looked at before.

SCHEMING AND PLOTTING

So you know what's up and you definitely want to observe tonight — time for more planning. If you're going to use your telescope, keep in mind that it needs a while to cool down (unless you store it outside, away from direct sunlight). Scope tubes hold heat, though it turns out that the worst problem is the objective (the main mirror or lens). If it's not at air temperature, you'll never get a great view of anything, because you'll be looking through wavering currents of warm air, generated by the mirror or lens, that are trapped inside the telescope. The view markedly improves about 30 minutes after the telescope is taken outside, but the larger the scope, the longer the cool-down time. It pays to set up early.

Consider the weather. Are clouds going to roll in and spoil your night? (Turn to page 47 for a brief look at meteorology.) Don't forget the Moon. When it rises and sets and how bright it will be can make or break an observing session. Where moonlight is concerned, a little goes a long way toward spoiling the night sky, so if you're going after faint targets, avoid the nights around full Moon.

If you can, have a nap in the afternoon. Even an hour or two of shut-eye will make

WHAT TO SEE

There are as many different ways of deciding which objects to see as there are observers trying to see them. You'll soon develop your own methods of determining what to look at, but here's one idea. If you've seen or read about something you would like to view, jot it down in a small notepad. Then find out when it will be conveniently placed in the sky for observing. If you use loose-leaf paper, you can easily organize a "Things to Observe" book by month, season, or whatever suits you best.

Keep in mind the limits of your telescope and observing site. Some objects may not reveal themselves in a city sky because of light pollution, or your scope's aperture may simply be too small to capture the fainter fuzzies that inhabit the heavens.

Kemble's Cascade NGC 1502 Camelopardalis
04h 07m 49s +62° 19' 54"
6.9 mag Open cluster

M81 NGC 3031 Ursa Major
09h 55m 32.9s +69° 03' 55"
M82 NGC 3034 Galaxy
09h 55m 50.7s +69° 40' 43"

M13 NGC 6205 Hercules
16h 41m 41.4s +36° 27' 36"
globular 5.8 mag

M92 NGC 6341
17h 17m 07.5s +43° 08' 11"
6.5 mag
M7 NGC 6475 Scorpius
17h 53m 51.1s −31° 47' 34"
3.3 mag Ptolemy's Cluster

Coathanger - Collinder 399 Vulpecula
19h 25m +20° 11'
3.6 mag between Vega and Altair

Blinking Planetary Cygnus NGC 6826
19h 44m 48.0s +50° 31' 31"
8.8 mag

Helix Nebula Aquarius NGC 7293
22h 29m 38.35s −20° 50' 13.2"
7.3 mag – very diffuse

Your observing list doesn't have to be a fancy computer printout. Even a handwritten page of ideas will suffice.

the difference between staying up late and still being productive or crashing early from fatigue.

Make sure that the objects you want to see will be well placed for viewing. This includes taking into account not only their position in the sky but also any nearby tall trees and buildings that may block your view.

VACATION UNDER THE STARS

Sometimes I like to sit and dream about my ideal astronomical vacation. Then every year I do it by going to the Texas Star Party. Star parties are great events to meet fellow enthusiasts, pick up observing tips and techniques, and look through some truly giant scopes. They're often held over a weekend, so you need use only a little of your precious vacation time to attend. I have more to say about star parties in Chapter 4.

More often than not, I end up squeezing a little astronomy into a regular vacation. It's not hard to do; see page 38 if you're thinking

Above: **It may look like an ordinary campsite, but look again. The presence of telescopes (most of them covered) scattered throughout the area tells you that it's the site of a star party.**

Left: **If camping under the stars isn't for you, consider staying at an astronomical bed and breakfast, like this one near Osoyoos, British Columbia, Canada, that's operated by world-famous astrophotographer Jack Newton and his wife, Alice. Here you have the run of the observatory (after a little training) or can get help taking great astro images. Best of all: no noisy neighbors up at the crack of dawn just as you're trying to get to sleep.**

about taking a telescope with you. If you're driving, consider stopping at a potential dark-sky site such as a national park. Even without a scope, an evening under a sky free of light pollution can be truly inspiring.

But now and again a particular celestial event such as a total eclipse of the Sun (which happens when the Moon completely hides the Sun for a few minutes) is the reason for a vacation. Whether you join a tour group — eclipse chasing has become a big business — or travel independently, you'll definitely experience a vacation to remember.

A total eclipse of the Sun is a spectacular sight (I've seen two) and well worth the journey required to view it.

Meteorology and You

When you become a novice astronomer, you also become a novice meteorologist, and *that's* something you won't read in any telescope manual! Taking on this extra hobby happens automatically because planning your observing session has to include trying to figure out what's up with the weather. While meteorology is a com-

DEW DROP IN

It's a clear night, you've been observing for a while . . . and now everything you look at through your scope seems dim. A quick check reveals a layer of dew coating your optics. (Dew appears on any object that becomes colder than the air's dew point. When this happens, your telescope starts to drip and a foggy mist forms on objective lenses, mirrors, and eyepieces.)

The options for fighting back begin with prevention. An extension on the objective end of your refractor or compound telescope (reflectors don't need this) will thwart the moisture. Such a *dewcap* can be purchased or made out of simple materials *(right)*. Don't forget one for the finder, and keep eyepieces in their case with the lid closed. You can buy a heated dewcap or anti-dew warming strips from many astronomy-gear suppliers; attach and activate the device at the start of the evening.

If dew is present, wiping it off doesn't help; the mist returns almost the moment you're done. Instead, use a hair dryer on its lowest setting to blow warm air across the affected surfaces until the dew vanishes.

plex subject, there are only a handful of weather-related issues I consider when I'm thinking about heading out for a night of observing.

During the day, I watch for contrails left by aircraft. If they're long and lingering, my scope usually stays inside because from experience I know the sky will be "mushy," high cirrus clouds are likely to form, and the night will be a frustrating challenge of hide and seek between celestial object and cloud. If the trails are nonexistent or very short, my scope and I go out.

A deep-blue daytime sky will likely find my scope outside waiting for darkness. But a pale-blue or washed-out sky means there's a lot of dust in the air, which will dim the sky and make it harder to find faint objects.

As a final post-sunset check, I look to see what the stars are doing. If they're twinkling very strongly, that's bad. Turn the page to "Twinkle, Twinkle" for the reasons why.

WEATHER WATCHING: HAPPY OR LOUSY

I know this is a very generalized view of an intricate subject, but a quick glance at a weather map can give me a sense of what to expect during the upcoming 24 to 48 hours.

High-pressure systems are usually associated with fair weather. When there's high pressure in your forecast, it's likely the sky will be mostly clear, with little or no precipitation. If you want to observe, look for a large *H* on the weather map either approaching or hovering over your area. For astronomers, *H* often means "Happy"!

Low-pressure regions typically produce stormy weather. On a weather map, the large *L* indicates the location of a low; for observers, that *L* frequently stands for "Lousy." The only consolation is that after a low passes, the sky usually clears.

If you hear that a front is coming your way, consider staying indoors to study your star charts (or read more of this book). Both warm and cold fronts are associated with unstable weather, includ-

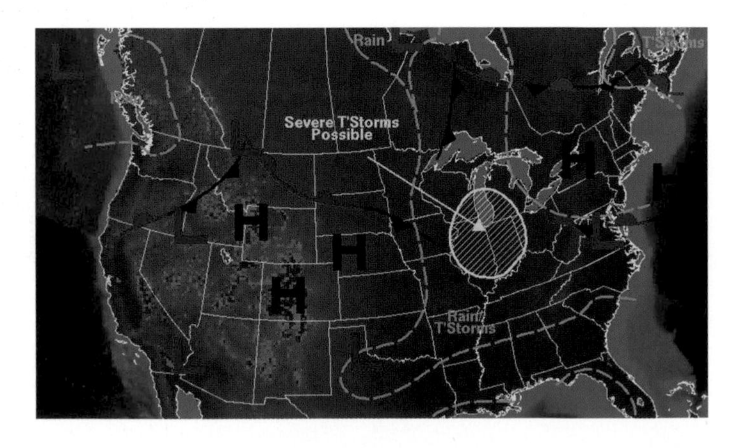

BASIC CLOUDS

There are three general cloud types — cumulus, cirrus, and stratus — but variations and combinations of these three basic types can make it difficult to identify the clouds you see.

Cumulus clouds look like big cotton puffs and are the "good" clouds because they usually go away after sundown.

Cirrus clouds are wispy high-level clouds. They generally occur in fair weather but often linger. If they do, you'll still be able to see the Moon and the bright planets, but that's about all you'll see.

Stratus clouds are horizontal, layered clouds and usually cover most of the sky. Keep your scope inside until they dissipate, because it may take a while for them to do so.

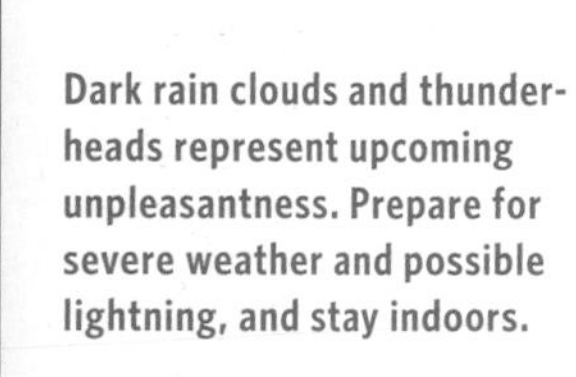

Dark rain clouds and thunderheads represent upcoming unpleasantness. Prepare for severe weather and possible lightning, and stay indoors.

TWINKLE, TWINKLE

"Twinkle, twinkle, little star" sounds visually appealing in a childhood song, but twinkling stars usually indicate that viewing conditions are not ideal. When the stars are twinkling like mad and you can't get them to stay focused in your scope, the atmosphere is turbulent and astronomers say the *seeing* is bad. Seeing often improves as the night wears on, but it's always poor near the horizon.

The clarity of the atmosphere is known as *transparency*. Dust, haze, and even pollutants can affect your view of celestial objects. If the sky isn't particularly transparent, you'll have a hard time finding faint objects, so your observing could be restricted to the Moon, planets, and bright double stars.

The maddening part about all this is that a clear sky doesn't guarantee good viewing. When seeing is good, transparency can be poor, and vice versa.

ing clouds and showers (or snow in the case of a cold front). But if a cold front has passed and a high-pressure region is moving in, get outside — observing conditions are often great in this situation.

Finally, check to see where the jet stream lies. It moves around (sometimes a lot), and if you're within a couple hundred miles of it, the *seeing* will probably be poor because the stream's winds cause the atmosphere to be very turbulent.

Chill Chasers

Even if you've spent only a couple of nights under the stars, you've probably figured out that astronomy is not an aerobic sport. Sure, it has its moments, particularly when you're lugging the telescope and all its accessories from the basement to the backyard or from your vehicle to an observing spot. But this activity eventually slows to a tortoise's pace once everything is assembled.

So unless you're using your counterweights as dumbbells or running laps between your scope and your house, stargazing is a

Here's the well-dressed astronomer ready for a winter-night's observing!

Even if you observe where the palm trees grow, it gets chilly at night. Make sure you dress appropriately.

mostly sedentary activity, interrupted by only the occasional need to push your telescope around. It's worse if you have a Go To telescope — then the only part of your body you'll be moving is the finger you need to punch some buttons! And no movement means you're not generating much heat, which means you'll likely get chilly even when the temperature isn't low.

Staying warm while stargazing is easy, provided you're prepared. This means dressing right and not worrying about how unfashionable you look. Comfort is an important (and often overlooked) subject, but believe me, you'll enjoy the experience of observing a lot more if you're warm.

Even balmy summer nights can get cool by 3:00 a.m. so be sure to have a jacket handy. A windbreaker or thin rainproof nylon jacket is a multi-purpose garment that wards off the chill, packs small, and sheds water. You won't be observing in the rain, but you might experience some overnight dew. And sudden summer thunderstorms have caught many a stargazer off-guard. (Of course we all know that in such instances the jacket isn't for you — it's to protect your optics until you can get them properly covered!)

CLOTHES MAKETH THE OBSERVER

Beyond the summer season, observing can be a cold passion. Every article or book I've read about staying warm during cold weather talks about layering. Here's how to apply that concept to observing.

The first trick isn't right out of the hat — it *is* the hat. A lot of heat escapes from your head, so keep it covered with a cap, hat, or scarf. If your jacket has a hood, use it.

Hands and feet are another problem area — once they're cold it's hard to satisfactorily warm them again. Fingerless gloves (with mittens over the top), mittens with fingers that fold back, or a combination of mittens over the top of thin silk gloves are great for observing. Not only do they keep your hands warm, but also you can temporarily expose your fingers to get a better feel for the knobs you're adjusting or the eyepiece you're changing. At the other end of your body, insulated boots (not running shoes) and a double layer of socks will keep your toes toasty. Be sure to wear socks that wick away (carry away) moisture because your feet can perspire even when it's cold.

Flannel or fleece-lined jeans and quilted shirts are useful when you observe during an in-between season when it's not cold enough to wear everything you own. Still, the arrival of winter usually means adding an inner layer of thermal underwear. Fortunately, you no longer have to waddle around looking like an abominable snowman just to stay warm; thermal underwear is now available in thin silk and silk-like fabrics. Again, look for material that will wick moisture away from your body. And naturally you'll be wearing a jacket that's suitable for the season.

Finally, don't just stand there . . . do something! If you're part of a group, walk around and visit your astronomy buddies to see what they're looking at and rev up your circulation to generate a little body heat at the same time. Observing by yourself? Then there's no one around to chuckle as you run in place or do some jumping jacks for a couple of minutes every half hour. If your fingers are cold, swing your arms in large circles to increase blood flow to your frigid digits. Of course if you're in your backyard, you can always go inside to warm up. Just remember to wear your dark-adaptation goggles (page 57) and don't stay indoors too long; becoming overly warm can lead to the chills once you're back outside.

WARMTH THROUGH CHEMISTRY

Hand-warmer packs are small chemical heaters that are activated by shaking and come in assorted sizes to accommodate pockets, gloves, and boots. They will keep your hands and feet toasty for about two to four hours. Most outdoor-supply stores carry them in winter; just be sure to follow the instructions on the package.

Unless you live in Florida and plan to observe in Alaska, most of the clothing you need for observing is probably already in your closet.

That blanket you spread on the ground to watch meteors in August will help keep you warm when you observe in January. If you sit for a while, wrap it around yourself to keep your body heat from escaping.

SNACK PACKING

While we tend to think of snacks as nothing more than a quick pick-me-up that'll help extend our stay under the stars, consuming the right stuff may make the difference between observing all night and fading in the dark. Eating warms you up because it takes energy to digest food and convert it to fuel for your body. But eat the right foods, because the wrong snacks can make you sluggish and sleepy. Vegetables and fruit (the non-messy kind) cut into small strips and packed in plastic bags make good snacks. Cheese is an easy take-along source of protein, and nuts and granola bars can be stashed for an energy boost. And I don't know how I'd ever have made it through some long observing nights without chocolate! Whatever you do, avoid goopy, syrupy, or sticky foods and anything that might crumble (potato chips, for instance), since I can guarantee that crumbs *will* find their way into your optics or equipment.

BETTER OBSERVING THROUGH HEALTHIER LIVING

No alcohol. In addition to making you less attentive, alcohol causes vasodilation, or widening of the blood vessels that can lead to increased heat loss. It also decreases the amount of oxygen your body receives, and since your eyes need a good supply of oxygen to work well at night, alcohol will impair your vision.

No smoking. Carbon monoxide exposure from inhaled cigarette smoke causes a reduction in oxygen to your body, which, as noted above, affects your eyes.

No fatty foods. These meals take your stomach a while to digest. Digestion ties up oxygen, and guess what? Low oxygen in the body reduces your observing prowess.

Most of the goodies you're likely to consume during an observing session are probably the same ones you'd take on a camping expedition.

> **COFFEE ANYONE?**
>
> Caffeine is a widely used stimulant that hits the nervous system soon after consumption, and its effects can linger for hours. Its main effect is, of course, increased wakefulness — it can perk you up. But it can also cause your blood vessels to constrict. If your extremities (fingers and toes) are particularly affected by the cold, you should keep your caffeine consumption low. And a high intake of caffeine could negatively affect your ability to see faint fuzzies. As always there's a middle ground; you just need to find yours.

Water is good, and it's almost impossible to drink too much. Even if you think you're not thirsty, drink some anyway — staying hydrated is one key to staying alert. Fruit juice is one of my favorite drinks. Sports drinks come in an array of flavors and have electrolytes that may give an extra boost. Hot chocolate, coffee, and tea are always welcome on cold nights, but see my note about caffeine at left. If you use a mug, make sure it's bottom heavy so it's less likely to spill and create a new nebula in your star atlas (or worse).

The Comfort Zone

Staying warm — being properly dressed, hydrated, and nourished — is only one aspect of being at ease while observing. There's another, but it's rarely discussed. Yet every time you go out for a prolonged session under the stars, it gets your attention. An aching back, tired feet, a stiff neck — observing can be hard on your body! That's why you need to get into what I call your comfort zone (others call this topic the ergonomics of observing). You won't observe for very long unless you're comfy. Fortunately, there are numerous ways to avoid turning stargazing into a painful experience.

COMFORT IS KEY

If you're going to be standing for long hours, consider using an anti-fatigue mat. These resilient foam pads will help reduce the strain on your knees, legs, and back. Some are available in interlocking squares that can be arranged to fit almost any viewing site. If you don't want to go that far, think about using cushiony inserts in your shoes; they'll help. And if you must kneel for any length of time, use gardener's kneepads or at least a rolled-up towel as padding under your knees to prevent rocks, sharp sticks, or prickly ground cover from poking holes in your kneecaps.

Left: **With an adjustable chair designed especially for use at the telescope, observing for long periods of time becomes much less painful.**

Above: **I find that a stool and a small table are useful accessories — and necessary for my observing comfort. The table keeps everything together and within easy reach, and even if I don't always need a stool to observe, it's nice to occasionally just sit and relax.**

Left: **When a large Dobsonian is aimed high, a proper stepladder is often required so an observer can reach the eyepiece. If you need one, don't go cheap. It's not worth saving a few bucks if the ladder collapses under you.**

But why stand (or kneel) when you can sit? Stools, chairs, and benches are essential observing tools if you want to stay out all night. Padded stadium seats offer soft spots for tired bottoms and may also provide extra warmth in winter. A musician's stool with a rotating seat will provide hours of comfortable sitting at the eyepiece, though its up-and-down motion is limited. If you want to go all out, there's a special adjustable astronomer's stool that makes viewing easy at any height. It folds flat, takes up little space for storage, and is easy to transport.

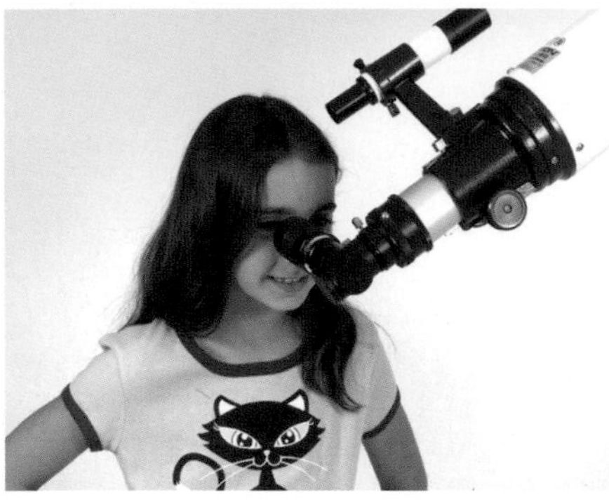

Above: **A star diagonal (on refractors and compound scopes) or a rotatable tube (for reflectors) allows different observers to adjust the position of the eyepiece to suit their height.**

Below: **Whether watching for meteors or satellites, exploring the constellations, or roaming the sky using binoculars, comfort and warmth are key, particularly when you're reclining. Of course, there's also the danger that you'll be so comfortable, you'll doze off and miss a fireball!**

It's difficult to truly observe an object if you're hunched over, standing on your tiptoes, or straining your neck in an awkward position just to get a peek through the eyepiece. Fortunately, the tubes of most equatorially mounted reflectors can be rotated to raise or lower the position of the eyepiece. Refractors and compound telescopes use a star diagonal, which can also be rotated. All you need do is remember that these adjustments are possible (though they also alter the orientation of your view).

If your reflector is aimed high, a step stool will likely be required to provide that needed boost so you can reach the eyepiece. (Of course, it can serve double duty as somewhere to sit when you need a break.) For tall scopes, particularly large Dobsonians aimed vertically, a ladder becomes an essential part of your observing kit. Just make sure that it's firmly planted and steady before you climb heavenward.

On those occasions when you find yourself skygazing without a telescope, why not lie down to look up? A thick quilt, unzipped sleeping bag, or large beach towel spread on the ground is perfect for lying on to meteor watch or spot bright satellites passing overhead. Better still is a reclining lounge chair; with its back support and handy armrests it makes sky surfing with binoculars seem effortless.

Seeing in the Dark

You've probably heard of night-vision goggles. But did you know that your eyes have pretty good night-vision capability if you give them a chance? It's called dark adaptation — the ability of our eyes to adjust to low levels of light. It takes approximately 20 to 30 minutes for the eye to become reasonably adapted to the dark, but if you're going after some really faint fuzzies (in other words, very dim galaxies or nebulas), your eyes will need nearly an hour of darkness to be ready for peak performance.

RED LIGHT

After waiting all that time for your eyes to become sensitive light-gathering instruments, it's really annoying to discover that a short burst of light is all it takes to destroy your dark adaptation — which means you have to start all over again! That's why one of the most important pieces of gear you'll need to stay adapted to the dark is a dim, red-light flashlight. Astronomers use them to read charts, write notes, make adjustments to their scope, and perform any other activity that requires light. But keep the red light dim while doing so; even bright red light can affect dark adaptation.

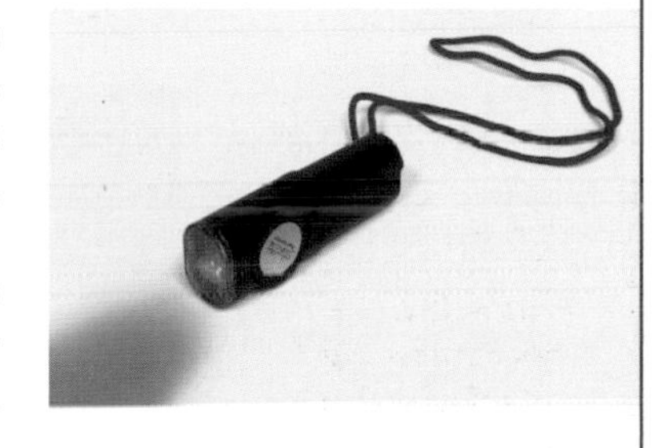

Many observers prefer flashlights that use red LEDs (light-emitting diodes) because they consume very little battery power and often have an adjustment wheel

Avoiding bright sunlight (or wearing heavy-duty wraparound sunglasses) during the day can help get your eyes ready for a night of observing. When you're out at night and need to go into a lit area, put on all-red glasses or goggles to help keep you dark adapted.

for brightness control (which also helps extend battery life). Make sure you get a "real" flashlight that uses two AAs or a 9-volt battery and not the tiny pushbutton type with a built-in battery — those won't last.

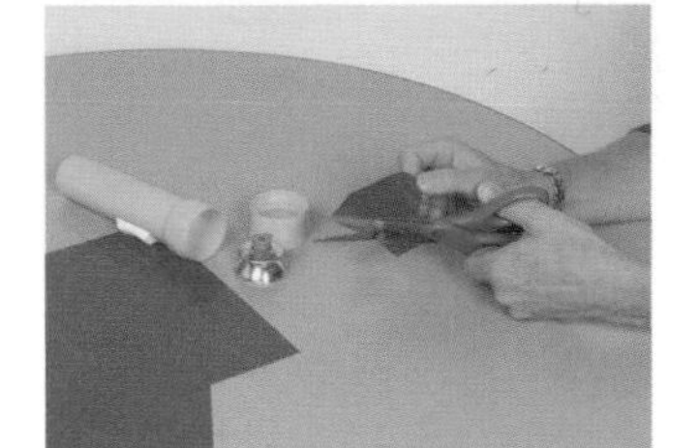

If you have a spare regular white-light flashlight, turn it into a red-light version by cutting a sheet of red plastic to fit inside or over the front of the light. You can also use a double layer of red cellophane fastened to the outside of the lens. Securely attach the cellophane or plastic, and err on the darker shade of red so the light coming out isn't too bright. Auto-parts stores have red taillight repair tape that also works well.

I've seen red nail polish painted on the lens (and even the bulb) of white-light flashlights as an emergency fix, but this is less than

DEFENDING THE DARK

Like a weed that sneaks into the garden, light pollution is slowly taking its toll on our nighttime harvest of stars. The loss of darkness isn't happening just in large cities (though it's particularly bad there). Rural areas that were once treasured dark-sky oases are feeling the pressure, too.

According to the International Dark-Sky Association (IDA), light pollution is any adverse effect of artificial light — including skyglow, glare, light trespass, light clutter, decreased visibility at night, and energy waste. You'd think that this description is fairly clear, but one person's light pollution is sometimes another's necessary illumination, though often the "necessary" illumination is badly done. Light that escapes skyward from poorly designed or installed streetlights, billboards, and security and decorative lights causes most of the skyglow that interrupts the darkness.

These pictures show the same section of sky above a suburban *(left)* and a rural location. Skyglow caused by city lights has little effect on the bright stars, but it washes out the diffuse light of the Milky Way.

Another type of light pollution is light trespass. This occurs when someone else's light shines on your property and causes a nuisance — almost everyone has experienced this in some form. There's also light glare due to excessive illumination, which is actually hazardous because it can reduce your ability to see.

What's to be done, and what can you do? The answers and solutions could fill a book, so instead I'll refer you to IDA's website (www.darksky.org). Here you'll find plenty of ammunition to take up the fight in defense of dark skies.

MOTION DETECTION

If you've ever watched a cat stalking its prey, you may have noticed that it moves its head in a barely perceptible side-to-side or up-and-down motion. If you do the same at the eyepiece (with your eye, not your entire head), this very subtle movement may make a faint object detectable. Tapping, bumping, or rocking the telescope's tube back and forth *very slightly* will also help you to notice detail or discover an object in the field of view that you didn't see before. Our eyes are automatically drawn to objects that move, and since you can't physically move a distant galaxy, the next best thing is to move the field of view it's in. A subtle nudge is sometimes all it takes to make an otherwise invisible object appear.

desirable since the polish will eventually burn and peel due to the heat of the bulb and result in uneven coverage. Beside, it smells awful as it burns!

AVERT YOUR VISION

The retina (the light-sensitive region at the back of the eye) contains two types of light receptors: cones and rods. The cones are packed into the center of the retina, work well under bright light, and let us see in color. The rods surround the cones, give us black-and-white vision, and work best in low light. Since the rods are located around the sides — at the periphery — of the cones, averting your gaze lets the maximum amount of light from dim celestial objects fall on them.

Try this on a dim star in your eyepiece. First, stare straight at it. It'll look very dim, maybe even disappear. Now shift your gaze so you're concentrating on a space toward one edge of the field. The star will reappear and may seem brighter than it did before. This is *averted vision*. It's like sneaking a peek or looking out of the corner

Since you're using primarily those black-and-white rods to observe with, don't expect the universe to appear vividly colored (no matter how large your telescope). Planets and some stars show limited hues in the eyepiece, but most deep-sky objects appear only as grayish glows. The colors in the Orion Nebula become obvious only in long time-exposure images; the left-hand photo has been altered to resemble a telescopic eyepiece view.

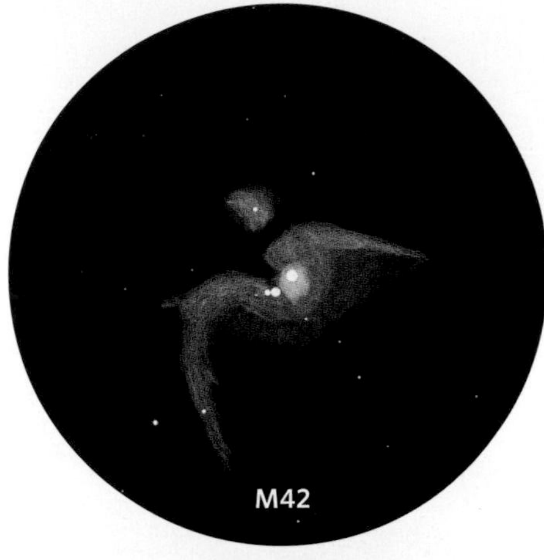

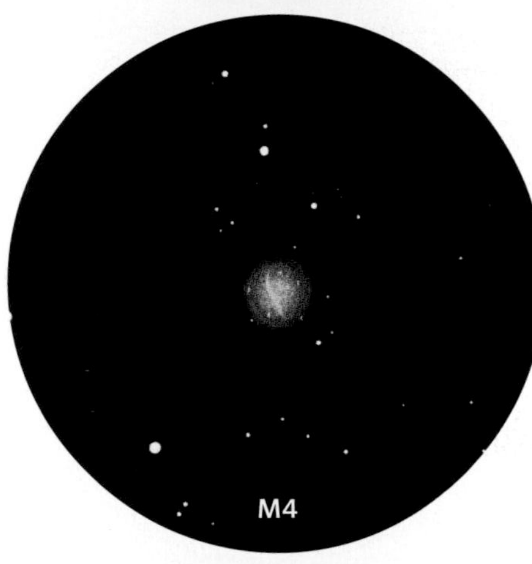

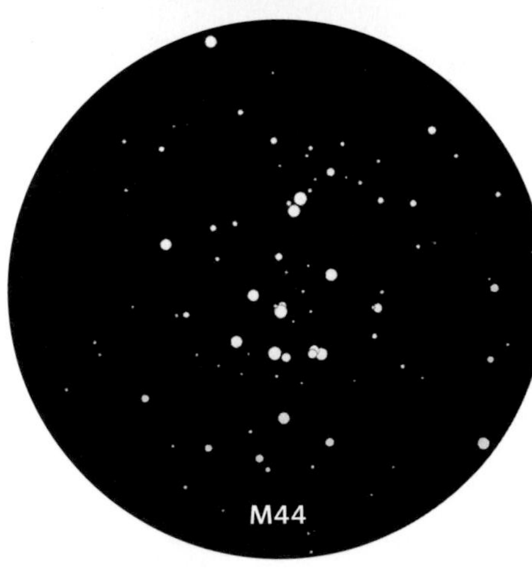

FINDING FAINT FUZZIES

Once you're in the dark, here are some tricks that will help you spot your faint quarry. The sketches (all by *Sky & Telescope*'s Tony Flanders) show what some deep-sky objects look like through a small telescope (north is up for all of them).

Breathe deep. Take a few deep breaths to rev up the oxygen level in your eyes before looking through the eyepiece. Increased oxygen helps your eyes perform better. And don't hold your breath while observing.

Use eye drops. Moist, comfortable eyes see better. (Caffeine, alcohol, and smoke can cause eye dryness and irritation for some people.)

Use filters. Some eyepiece filters are specifically designed to enhance contrast. Read more about filters on page 31.

Use motion. A gentle nudge of your scope so the field of view gently moves back and forth is sometimes enough to make an otherwise invisible object appear.

Look straight up. The darkest part of the sky is overhead. You can't always look in that direction, but at least try to observe your target when it's highest in the sky.

Choose your target wisely. Depending on the condition of the sky (including the amount of light pollution present), an extended nebula may be impossible to detect, while a compact cluster of stars pops into view. In general, small, high-contrast sky objects are much easier to spot than larger diffuse ones.

Observe during new Moon. Moonlight equals light pollution for deep-sky observers. It reduces contrast in the sky and can be detrimental to keeping your dark adaptation (especially if you look at the Moon).

Observe after a thunderstorm. Dust and water sometimes settle out of the atmosphere after a storm, resulting in improved transparency (though the seeing may be poor).

Don't wear yourself out. Fatigue will cause you to be less alert and less able to detect faint objects. Take frequent breaks and give your eyes (and feet) a rest.

Be careful of stray light. If you're observing where car lights may be a problem, close your eyes or turn your head away when you hear approaching traffic.

Be patient. If at first you don't see your target (and you're sure the scope is aimed correctly), keep looking, and use averted vision. Prolonged scrutiny usually starts to bring out details.

Practice. Your observing skills will improve every time you look through an eyepiece.

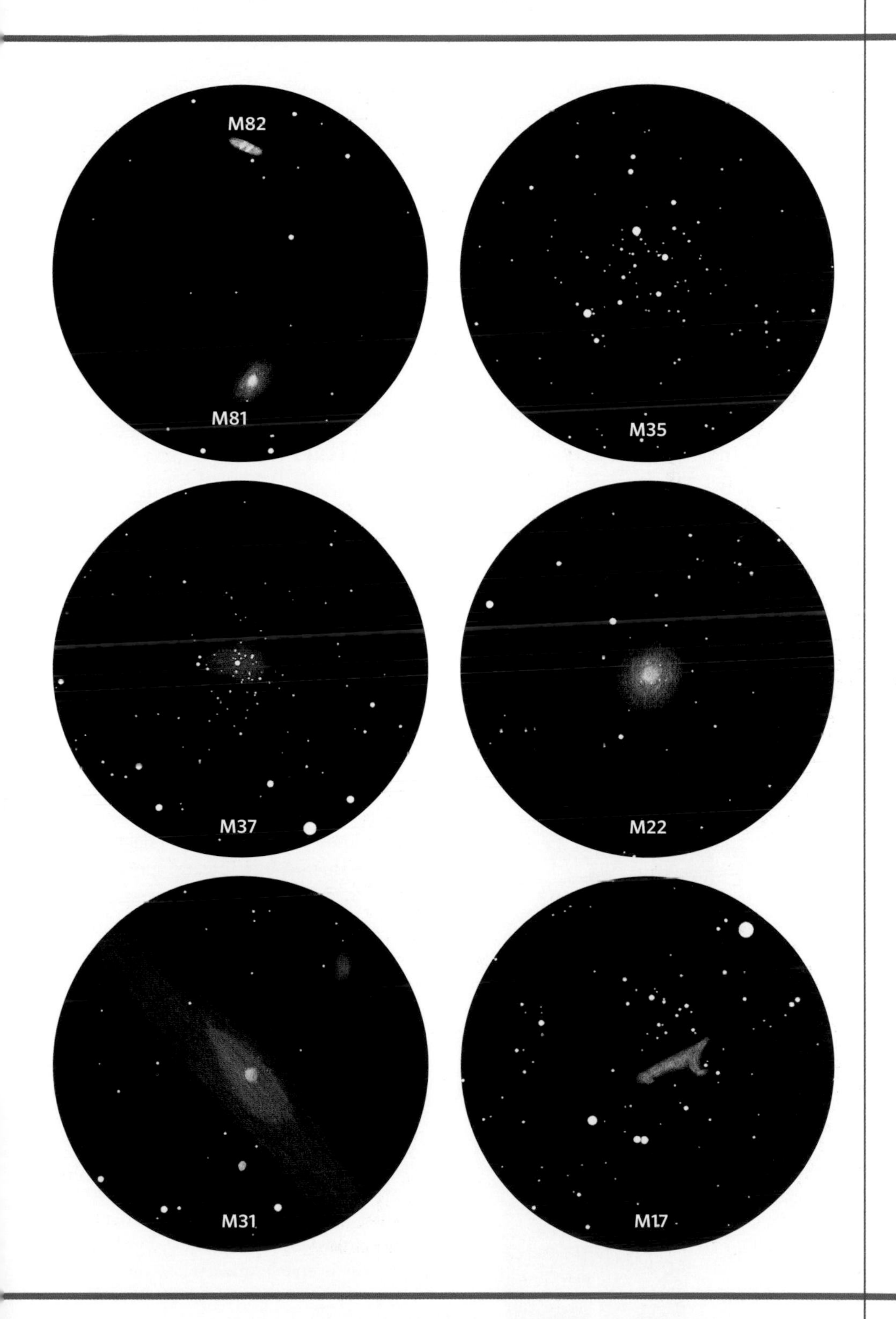
M82
M81
M35
M37
M22
M31
M17

of your eye at something. You're seeing it without looking at it. Sounds crazy, but it works. Averted vision is not magic, but you'll use it all the time when going after faint fuzzies.

After trying this technique a few times, you'll likely notice that you can see an object better by looking to one side of it rather than to the other. For example, use averted vision to look first to the left, then to the right of a faint object. Does the object seem brighter when you stare in one direction rather than the other? If so, you've found your sweet spot; gazing toward the opposite side reveals your blind spot. For most people, that sweet spot lies in the direction of their nose! Also try staring above or below the object; you may find that averting your eye upward works just as well. After a little practice you'll automatically look on the sweet side.

Sky Measurements

Beginners often have trouble describing distances in the sky because it doesn't work using familiar linear measurements like inches or millimeters. What looks like several inches in the sky to one person may seem like a foot or two to another. Instead, astronomers use angular measurement — degrees — to describe how far apart things seem. If someone says two stars are 15° apart, it simply means that the angle between the two stars and their eye measures 15°. You don't need expensive equipment to measure these separations; in many cases your hand will do.

10°

WIDE ANGLES

Hold out your fist at arm's length and sight past it. The width of your fist from one side to the other covers about 10° degrees of sky. Now open your hand and spread your fingers wide. The distance between the tip of your thumb and the end of your little finger is roughly 20°. Your little finger covers 1° of sky — that's twice the diameter of the Sun or Moon.

NARROW ANGLES

It's also helpful to know how much sky you can see through any particular eyepiece. There are ways to calculate this, but here's a simple trick that'll at least give you a feel for what a low-power eyepiece is showing.

Take your telescope outside when the first-quarter Moon is visible. Use your lowest-power eyepiece and center the Moon

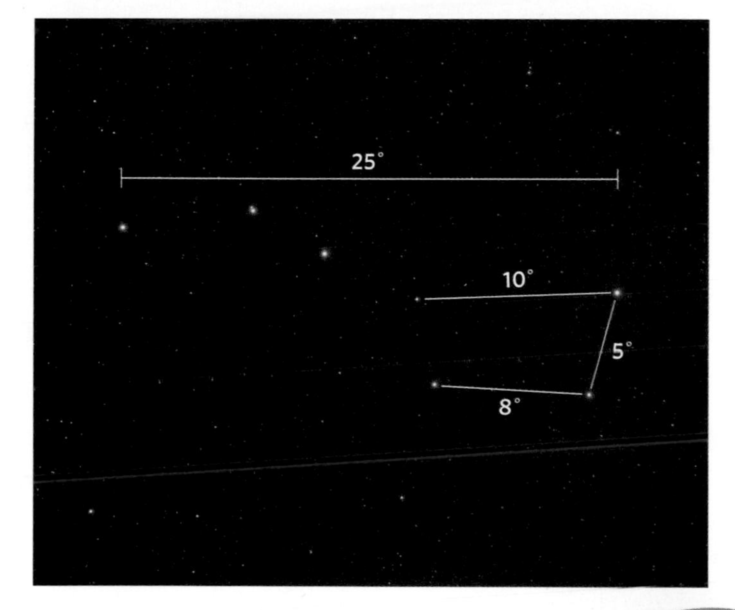

The Big Dipper is an easily recognized pattern for almost all Northern Hemisphere stargazers. Use it to help estimate angles in the sky. At the end of the Bowl the pointer stars (so-called because they point to the North Star) are 5° apart. If they fit snugly within your finder-scope's field of view, it means your finder shows 5° of sky.

in it. Remember, the Moon is ½° in diameter (that's 30 arcseconds). Does it fill your eyepiece? If so, your eyepiece shows ½° of sky. If there's lots of space around the Moon, move it to one edge of the eyepiece and try to estimate how many "Moons" could stretch across your field of view. If it looks to be two Moons wide, then you're seeing a 1° field.

Do the Star-Hop

No, star-hopping isn't a new dance. It's actually an old one, and it takes place in the night sky! Simply put, a star-hop uses stars and star patterns as celestial guideposts, which direct you toward a faint object that might be difficult (or impossible) to see in your finderscope. Star-hopping is a simple technique that will help you find your way to celestial targets and, in the process, may lead you to sights you'd otherwise miss. In this age of Go-To technology, star-hopping is becoming a lost art, which is why I think it qualifies as a secret! Star-hopping is your ticket to many nights of adventure exploring the endless sea of stellar treasures overhead. Frankly, I think Go To owners are the poorer for letting their scope do all the work, though, as I've said before, Go To's do have their place.

You've probably already star-hopped and not even realized it. Have you ever used the pointer stars in the Big Dipper to find Polaris? Then you've star-hopped — from the Big Dipper to Polaris in one go. But most hops are a little more complex than this.

GETTING STARTED

The sky is big, and the thought of trying to find some tiny star cluster or faint galaxy can be overwhelming. It's easy to get lost without a set of good star charts and an observing plan. You also need to have a basic grasp of what's up there — the bright stars and constellations. If you can't identify these things in your sky (whether it's a light-polluted city sky or a dark rural oasis), you're not ready to hop.

Once you're comfortable under the stars, it's time to pick up a proper star atlas. (The one I recommend is *Sky Atlas 2000;* it's listed in Chapter 5.) The charts may look complex, but look again. You should be able to identify the bright stars you can see in your sky and link them to the constellations in which they reside. Do be careful of something that could cause you some confusion: directions. In a star atlas, north is up, but north is *always* toward Polaris, the North Star. And unlike an earthly road map, east is to the left on a star chart (when north is up). Keep these two things straight, and you're well on your way to navigating the sky.

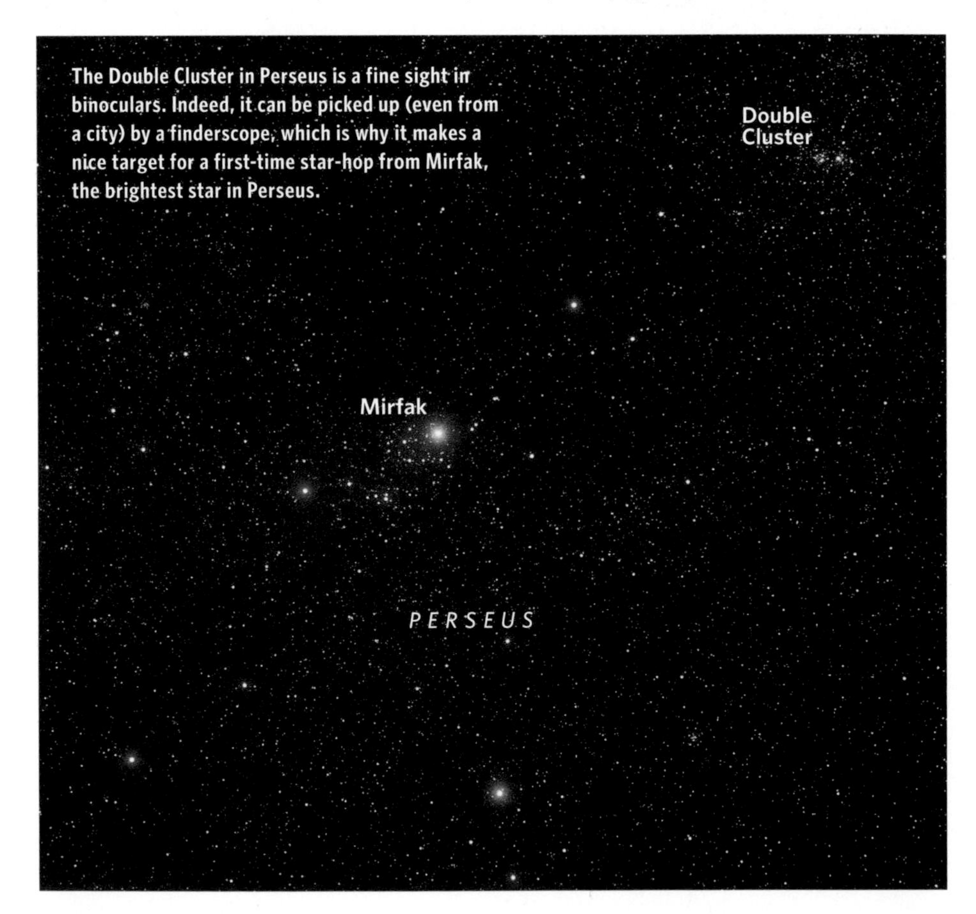

The Double Cluster in Perseus is a fine sight in binoculars. Indeed, it can be picked up (even from a city) by a finderscope, which is why it makes a nice target for a first-time star-hop from Mirfak, the brightest star in Perseus.

PLOTTING A COURSE

Before you plan your first star-hop, there's one more thing to do. Make a couple of wire loops that match the field of view for your finder and star-hopping eyepiece to overlay your charts. That's why the previous section — "Sky Measurements" — is so important. You need to figure out how much of the sky your finderscope and lowest-power eyepiece can take in. I use a 5° circle, a fairly typical field for a finder, and a 1° circle to represent the view through a low-power eyepiece.

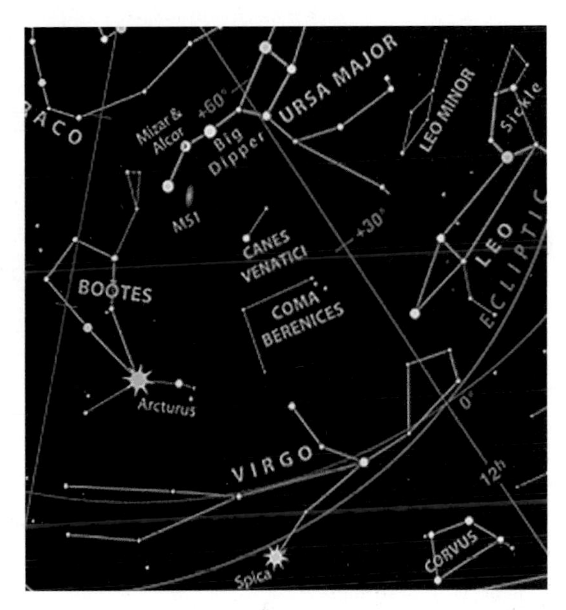

To star-hop to faint objects, you need a detailed star chart. A planisphere (right) won't show you where in Coma Berenices to look for deep-sky objects, even if you know there are some near Gamma (γ) Coma Berenices, the star that's at the right end of the horizontal line above the word COMA. The *Sky Atlas 2000* chart segment *(below)*, which shows a 10° × 10° slice of the sky, is centered on Gamma (circled) and reveals the location of numerous galaxies (in red).

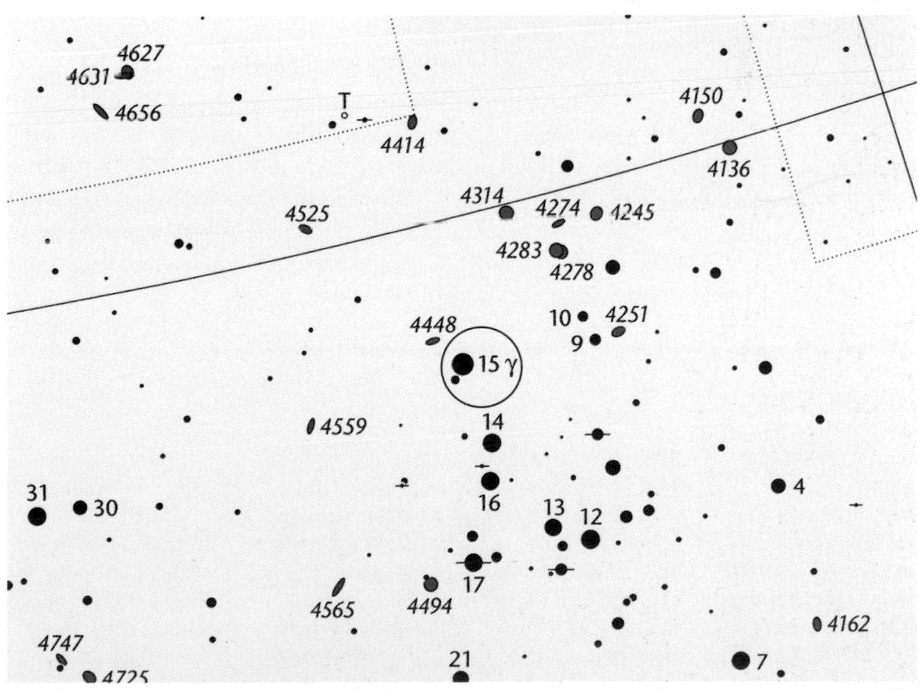

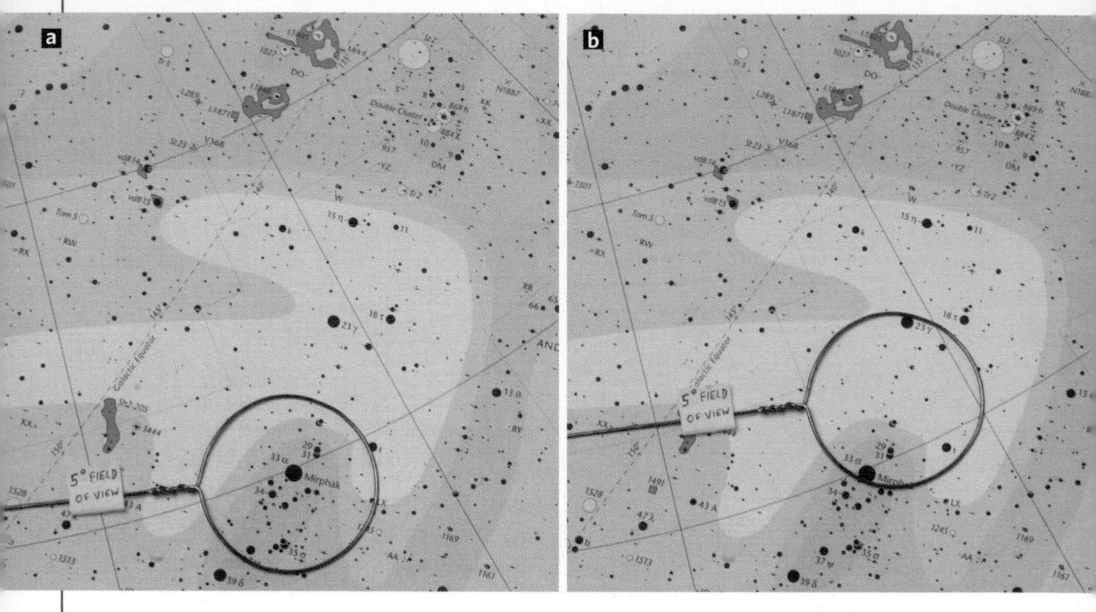

Start with an easy hop — perhaps from Mirfak, the brightest star in Perseus, to the famous Double Cluster of stars in Perseus. Here's how. Center Mirfak in the 5° ring (a). There's no other bright star within the field, but if you slide the ring so Mirfak goes to the ring's lower-left edge, you'll see that the star called Gamma (γ) appears at the upper-right edge (which, in the sky, is northwest of Mirfak) (b). Continue to slide the ring in

For your first hop, pick an area of the sky with which you're at least a little familiar. Circle a few targets you want to try hopping to — perhaps one of those Messier (M) objects you've read about or maybe a double star that's marked on the chart. Center the 5° circle on a bright star in the vicinity of your target, then gradually move the ring toward your target in increments of roughly 5°. Along the way, create some easy-to-recognize patterns (by connecting the stars) that'll help guide you toward your intended target. Form triangles, rectangles, diamonds, or whatever it takes to make the path recognizable. I like to make triangular patterns out of small groups of stars to hop across the sky.

MISSING STARS

Sooner or later you're going to encounter stars that appear in your eyepiece but not on your atlas — particularly if you're observing from a dark-sky site. Rest assured that the atlas makers didn't forget, or deliberately omit, some stars. All star atlases have a limiting magnitude below which objects are not plotted. Since most non-stellar objects are pretty faint, you'll find that many atlases have a magnitude limit that's brighter for stars (say magnitude 7 or 8) than for galaxies and nebulas (to magnitude 10 or 11). So it's very possible that you'll star-hop to a 10th-magnitude galaxy and see several faint stars nearby that don't appear on your atlas.

On the bright end of the scale, planets aren't plotted either, because they move. So if you discover a bright extra star in the sky, it may be a planet.

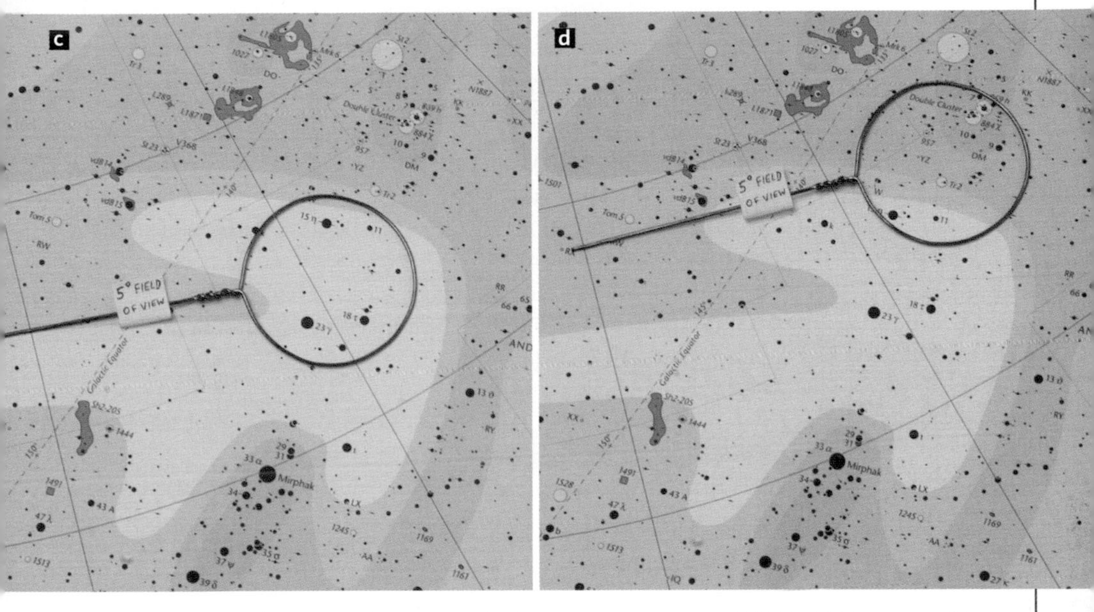

the same direction, and as Gamma drops down, a third star appears in the upper right: Eta (c). Now you need to change direction a bit. Put Eta (η) at the left (eastern) edge of the ring. At the upper-right (northwestern) edge of the 5° field, and therefore at the edge of your finder, lies the Double Cluster (d). Center the cluster by moving the ring/finder directly west by half its field, and the cluster should be visible in the eyepiece.

If your star charts are laminated and printed with white stars on a black background, use a white grease pencil to plot your hop and sketch those little markers. Afterward, the grease marks can be wiped off the chart, and you're ready to find another target. If you've got regular paper charts, lightly mark your path with a pencil and use the same trick of making small patterns with stars.

Once the 5° ring is resting over your target, study the field carefully. Overlay your eyepiece ring (the 1° circle) — you'll be amazed at how little you're going to see through your eyepiece. Once you know what to expect at the eyepiece and know which way to gently move the scope to bring your target into view, head outside and try the star-hop.

One last thing. Depending on the type of finder you have, the view through it might not match what your chart shows. See the illustration on page 37 to determine how your finder shows the sky. You might need to do some mental gymnastics to get the view through your finder to match what you see on your chart.

Star-hopping takes time and patience, and it's particularly challenging in a city sky. Here you may have to star-hop using your low-power eyepiece instead of your finderscope because your finder may be unable to see enough stars to make it work. Still, if you persevere you'll have the satisfaction of truly knowing the sky and discovering its many secret sights.

In Your Own Words

It's a high-tech world out there. So you may laugh when I tell you that Galileo — the famous Italian astronomer who lived 400 years ago — and I have something in common in addition to a passion for exploring the universe through a telescope. He kept an observing journal and I do, too — you know, the old-fashioned kind in which you write and sketch to record your observing experiences. It's a fun and easy way to track your nightly excursions among the stars.

Of course, there's nothing wrong with using a camera or webcam to record your observations — it's not an either/or proposition. But I find that the camera doesn't capture exactly the same scene that I observe through the eyepiece. It's important to me that I record what I see, so I sketch and write.

I've kept an observing journal in one form or another for more than two decades, and that's quite a collection of observations. But they don't just sit on a shelf. I often enjoy flipping through my journals and reading old entries to see how much I've changed as an observer and to revisit some cherished moments. In some ways a journal can be like a school yearbook. It reminds you of the experiences, people, and places of your past, and it refreshes memories that otherwise might become blurry in the mists of time.

I do try to take my own advice! Here I've settled in for an evening's observing session, ready to make notes in my ever-present astronomical journal. Recording what I see has definitely improved my observing skills.

THE PERSONAL TOUCH

Journal keeping has no hard and fast rules. It's up to you to decide what you want to record. You can just briefly note the date, the object looked at, and where you were when you viewed it. Or maybe you'll want to scribble longer entries with more information such as the time, sky conditions, telescope and eyepieces used, and a detailed description of the target — possibly including a sketch.

Here's my view of Comet NEAT as seen from Albuquerque, New Mexico, at 9:50 p.m. on May 11, 2004. I used a 3-inch refractor with a 19-mm eyepiece.

TSP-2006 April 22
Saturday night 9:51
nice ISS pass through the Davis Mountains. It appeared just below Orion and skimmed over the mountains to the west. Bats were out echo-locating moths drawn to the Motel lights.

Betelgeuse
• Mars
Path of ISS

Pass ended just behind north roof corner of house on Mountain.

As you can see, sometimes the sketches are pretty rough, but it doesn't matter — they reflect my observation and serve as a visual reminder of what I saw.

March 8, 2005 4:30 am
Omega Centauri - through the Kitchen window!
My first view of it this year was not planned, it just happened when I woke up to put another log on the fire and get a drink of water. While I was standing at the kitchen sink drinking and looking out admiring how bright Jupiter was, I noticed a fuzzy patch just barely above the tree line toward the southeast. Now I'm curious to see if it is truly Omega or if my still half asleep eyes are playing tricks on me. They are not! Through 8x56 binos the fuzzy patch through the window IS Omega. Now I want to see it better, but it's too cold for me to take the scope outside and too early.

Sometimes I write only a comment, which can bring back memories that are just as wonderful as if a photograph accompanied it.

Stargazing is a personal affair that changes with time, equipment, and experience. Journal writing is personal too, and that's all the more reason for keeping one. Over time, a journal can show how your approach to observing has evolved.

A logbook or diary can also help improve your observing skills. When you write down what you see, you start paying closer attention to your target. At first glance, most nebulas and galaxies just look like dim fuzzy blobs. But all dim fuzzies are not created equal. Is the object round or elongated? Does it have color? Is it brighter on one side or the other? Which side? Was it easy to find?

Don't be afraid to include your stargazing companions in your notes. Personal anecdotes about observing experiences add nuances that bring descriptions to life. One of my favorite journal entries is about stargazing with friends at a state park when a rather bold and noisy raccoon decided to visit and raid our snacks. While the critter wasn't an astronomical target, the shrieking of one of my companions made the evening memorable.

SIMPLE REQUIREMENTS

You don't need a telescope to keep a journal. Record naked-eye views from your backyard or from any other place you happen to be. Write down what you see in binoculars. Take your journal to star parties and make notes about the stellar wonders viewed through other

LOW TO HIGH TECH

A simple spiral notebook that opens flat can be used for your observations and notes. Another low-tech option is a three-ring binder; its advantage is that you don't have to take the whole thing with you to make an entry at the telescope. The binder itself is big and bulky, but a few loose-leaf pages are easily transportable.

At the other end of the scale are digital cameras, webcams, video recorders, and all manner of electronic recording devices including a PDA (personal digital assistant). You can also create a weblog and post your observations online. Of all these electronic marvels, I must admit that a digital tape recorder for making observations at the eyepiece is particularly appealing, because you don't have to look away from the eyepiece in order to "write" your notes. Even better, the recorder allows you to keep the use of light to a minimum.

people's scopes. This will give you some insight into how different the same target might appear through a variety of instruments.

Adding sketches in your journal can greatly enhance your entries. Drawings can make written descriptions clearer. You don't have to be a Leonardo da Vinci to make a sketch. Drawings can be as simple as those by Galileo, who drew a small circle to represent Jupiter and small Xs alongside to indicate where he saw the Jovian moons. But if you have the artistic skill, try to capture as much detail as possible. To keep fresh pencil sketches from smearing, a quick spritz of hair spray will help prevent smudging.

If you have a digital camera, by all means take snapshots of Moon-planet gatherings or other sky scenes; you can print them out later and add them to your journal.

JOURNAL EVOLUTION

Currently I record my observations and sketches in a small blank sketchbook. But when I first started, I had a copy of *Peterson Field Guides: Stars and Planets* that became full of marked-up margins and bulged with unorganized yellow sticky notes. My very first entry reads "Orion Nebula, Nov. 1988, the Best! From the banks of Bastrop Bayou." It's enthusiastic but brief, and I wonder now just how good Orion looked.

After a few years that guide was a cluttered mess, and it was clear that I needed a better system. I switched to preprinted forms

For an observation that's really important to me, I'll take my time with the initial sketch and later on improve the look of certain non-essential components (like the foreground).

The techno-wizards will scoff at sketching, but if you enjoy it (regardless of whether you have an aptitude for drawing), a sketch that compliments your written notes is a wonderful addition to your sky diary.

in a three-ring notebook but ultimately found it too bulky. The forms did give me a template and the discipline to record more detail and in-depth descriptions of my observations.

You can keep your records in whatever is comfortable, convenient, or at hand, but I suggest something portable and small enough to tuck in with your observing gear. That way it's always with you. I also like to use a plastic mechanical pencil because it's cheap, needs no sharpening, and is ideal for making quick sketches. I stay away from pens because ink freezes or gets sluggish in the cold.

Whatever method you choose, I urge you to start an observing journal now. I know this sounds a bit corny, but 10 years from now, you'll be glad you did.

What's the Hurry?

Imagine you're driving past a sandy white beach. You stop, get out of your car, and stroll along the shore. You notice that the monochromatic "white" sand is actually a mosaic of subtly colored individual grains sparkling in the sun. Scattered throughout are tiny seashells and fragments of driftwood. Small shore birds scurry through the shallow waves searching for marooned morsels of food. Bits of stranded seaweed glisten in sunlight.

None of this was visible as you sped past — it looked like every other white-sand beach you'd ever seen. The subtle nuances of color, texture, and activity could be appreciated only by pausing for a longer look.

PATIENCE PAYS

Slow down when looking at the sky; the rewards are as plentiful as those found on that beach. Your first glimpse of Jupiter through the eyepiece may show a small, round disk with a couple of faint streaks. But don't start thinking about the next thing on your list — look again. Are those dark streaks (Jupiter's belts) completely featureless, or can you tease out a little detail? Are there more than two belts? You've heard of the Great Red Spot; can you find it? Is there a tiny black dot visible on the planet's face? (If so, that's the shadow of one of Jupiter's four largest moons. See if you can find a nearby bright dot — that would be the shadow-casting moon). Look for the Jovian satellites on either side of Jupiter. Is one just starting to peek from behind the planet's limb? You'll miss a lot if you don't take the time to carefully examine what's right in front of your eyes.

Or perhaps you've found a cluster of stars that, at first glance, looks a lot like all the other star clusters you've already seen. Look again. How dense is the cluster? Is it symmetrical? Does one particular star in the field have a slight tinge of color to it? Is there something of interest in the cluster's immediate vicinity?

Unless you're participating in a race to find a certain number of objects during a single night, slow down, take your time, and be patient. The point of observing is to observe!

It doesn't matter whether you're looking without optical aid, using binoculars, or peering into a telescope's eyepiece — slow-

The star clouds of the Milky Way offer something for every type of instrument — from viewing with the naked-eye to high-power telescopic observing.

Some objects, like the Beehive Cluster in Cancer, are best seen in binoculars.

EXPERIENCE COUNTS

Observe as often as you can. It's really no secret that the more often you do something, the better you become at it. "Use it or lose it" applies to your observing skills, too!

ing down will allow you to notice more. And by using some of the tricks I've mentioned previously (like averted vision and being properly dark adapted), you'll see sights that may have escaped your notice in earlier attempts. The longer you look, the more you will see.

If you just can't seem to slow down, start an observing journal. The simple act of recording your observations will cause you to take many looks at the object in question to ensure that your description matches what you're seeing.

START LOW

Less is more. Okay, I agree — it's a nonsensical phrase. But when it comes to observing, I believe it. It's hard to beat a naked-eye view of the Milky Way glowing overhead in a dark rural site. Every time I attend a star party, the appearance of the Milky Way on the first night takes my breath away.

When you start out with no (or very low) magnification, it's easier to find your way around the sky and appreciate the big picture. Less magnification means a wider field of view — an important consideration if you're star-hopping. Examine the Pleiades (a beautiful open star cluster in Taurus, the Bull) at low and high power. A low-power view through binoculars reveals the elegance of the entire cluster. A high-power telescopic view shows only the individual starry components.

Left: **Sometimes you need a telescope and reasonably high power to explore celestial sights like the Whirlpool Galaxy — also known as Messier 51 — in Ursa Major, the Great Bear.**

Below: **One set of objects that will usually absorb high magnification is the planets. The detail that's always present in the Jovian cloudtops makes Jupiter a particularly pleasing target.**

Of course high magnification has its place, particularly when you're observing the planets. All I'm saying is don't rush in. Keep in mind that when higher magnification is used, the field of view shrinks and everything appears dimmer. Sometimes sky conditions simply won't support a lot of magnification. A sharp low-power image is better than a fuzzy high-power one. Experiment with two or three eyepieces to determine the level of magnification that's best for a particular target on any given night. Start low, work your way up, and when the view turns ugly, step back down to the previous level of magnification and accept what it gives you.

Finally, with low magnification you have a better walk-away factor (as I call it). If your telescope doesn't track the sky with a motor, low power allows you more time to study the object before it drifts out of view and gives you time to walk away from your telescope (for whatever reason) without the object drifting too far out of the field of view. Eventually, you'll instinctively know how far you need to swing the scope to reacquire the target based on how long you've been away from your scope.

The Observing *Experience*

When you're just starting out, finding your way around the night sky can be frustrating, aggravating, and wearisome, and it can even end in defeat. Believe me, I know. I taught myself the sky the hard way using compressed star charts in a pocket-size guide with a department-store scope on wobbly legs. Perseverance paid off and I'm glad I kept at it, but it's not a path I recommend. Instead, I suggest joining an astronomy club and attending star parties. In both cases you'll be rubbing elbows with people who share your interest and can help accelerate your discovery of the sky.

Join the Club

An astronomy club is your boarding pass to the stars. It's probably one of the best investments of time (and eventually a little money) a newcomer can make to learn more about this satisfying hobby. The advantages of being a member vary among clubs (and no two clubs are exactly alike), but they all offer the camaraderie of people with a common enthusiasm. All have members ranging from novice to expert, many of whom are willing and eager to share their knowledge with you.

Best of all, you don't need to join right away. Go to a meeting or two as a guest, ask some questions, attend one of the club's star parties — in other words, check them out first. And best of all: you don't need to own a scope to belong.

NOT SURE?

If you're the shy type and would rather test the waters without getting wet, check out your local astronomy club online. Most now have websites where you can find details about when, where, and how often the meetings occur, what activities take place, what type of loaner equipment is available, and if there's a dark-sky site or an observatory. Some clubs even publish their newsletters online.

SO WHAT GOES ON?

Most clubs have monthly meetings, and some present novice programs before the main program begins. Hands-on workshops and how-to programs may be offered on topics ranging from building telescopes and cleaning optics to astrophotography and image processing. Lecture-style presentations are also part of the mix and run the gamut from club members describing their activities to guest speakers discussing the latest results in astronomical research.

Celebrities always draw a crowd, though not every club meeting (or star party) features a guest speaker like NASA astronaut John Grunsfeld.

Astronomy clubs can also be the perfect place to find a support group or even a mentor. Yes, you can learn the sky on your own, or let your Go To telescope show it to you, but there's nothing like some quality one-on-one time to smooth your transition from raw novice to amateur astronomer. Even if the club doesn't have a mentoring program, it usually has the equivalent of a novice support group where you can have your questions answered (and get answers to questions you never thought to ask).

Besides regular meetings, many clubs have members-only observing sessions. Sometimes it's informal — one night a month a few members head out to a site with their telescopes to look at . . . whatever they want to look at. There may be the occasional group outing with a purpose — to observe a meteor shower, eclipse, or other special astronomical event. And many clubs hold annual dark-sky observing sessions. Many are just for members, but in other cases the club invites the world to drop by (see "Star Parties" on page 81). And let me say it again: you don't need a telescope to join in these activities. In fact, these observing sessions are the perfect opportunity to learn all about telescopes (and receive plenty of advice) if you're thinking of buying one.

Whether at a regular meeting or a star party such as Stellafane (above), presentations on all aspects of astronomy — observational and scientific — are a staple of club activities.

Located about 50 miles south of Albuquerque, the General Nathan Twining Observatory is available for use by Albuquerque Astronomical Society members. The facility includes several telescopes, observing pads, a kitchenette, and a separate building that is a combination classroom, bunkhouse, and warm-up area.

MEMBERSHIP HAS ITS PRIVILEGES

But there's more! Many organizations have a library of books, magazines, videos, and star charts that members can check out. Some clubs own binoculars, telescopes, eyepieces, and even imaging equipment, and a few also have a permanent observatory located at a reasonably dark site — all are available to members (though sometimes for an additional fee). Clubs in this situation offer training programs for new users that include one-on-one field instruction with the equipment.

Once you become involved, you'll discover that most astronomy clubs contain subgroups. These are special-interest groups (SIGs), which cater to specific astronomical activities that have become a passion for some members. These groups may include amateur telescope making, observing occultations, asteroid and supernova hunting, eclipse chasing, CCD imaging, astrophotography, and many more. Think of a SIG as a club within a club that

FINDING A CLUB

These days, the first place to start is the Web (see page 87). If you can't find a listing, contact your local planetarium or science center, or ask at the astronomy (or physics) department of a nearby college or university. Don't ignore your local high school; it may already have a small astronomy group, and the teachers involved might know of a larger club close by. Some major cities actually have multiple astronomy clubs scattered throughout their suburbs.

If you can't find a club in your area, perhaps you can persuade a few local enthusiasts to help you start one. But who? And how? Well, go back to the planetarium, science center, high school, or local college/university and meet with some of the staff connected with astronomy, space science, or physics. You might be surprised to find others who'd be interested in starting an astronomy club — if only *somebody* would take the lead. You can be that catalyst. And even if the new club consists of only a handful of people, you now have an astronomy support group as well as new friends who share an interest.

Above: **Many clubs hold regular sidewalk-astronomy sessions in their community.**

Right: **If you like building things and want to do something that most amateur astronomers don't do any more, why not make your own telescope? An astronomy club is the perfect venue for finding other amateurs with the experience to help you achieve your goal.**

offers a more intense exploration of a particular subject. After a while, if you discover some aspect of astronomy that you're passionate about, you may end up starting your own SIG!

Public outreach is another important activity. In fact, your first encounter with an astronomical organization could be through one of its outreach activities. Giving talks at schools, hosting skywatching parties in parks or on neighborhood sidewalks, promoting special astronomical events, participating in Astronomy Day, and encouraging communities to fight light pollution are just a few of the many ways club members interact with the public.

ASTRONOMY DAY

Astronomy Day occurs annually between mid-April and mid-May on a Saturday near or before the first-quarter Moon. Its theme is "bringing astronomy to the people." Astronomy clubs, science museums, observatories, universities, planetariums, libraries, and nature centers host special events (usually based on observing), many of which take place at non-astronomical sites like shopping malls, parks, and the sidewalks of urban centers. So if you're trying to find your local astronomy club, keep an eye out for Astronomy Day, attend some of the events, and learn who's behind them.

Party With the Stars

There's quite a difference between reading about observing and actually observing. I've already described the benefits of joining an astronomy club. Attending a star party adds yet another layer to the whole experience of being an amateur astronomer.

Star parties are not just a North American pastime — they're held by clubs around the world. Some are huge, weeklong gatherings that attract hundreds of amateurs from many countries. Others are smaller, more intimate events occurring over a weekend, with most participants being members of the local club that puts it on. The majority take place during the summer, and in the US there are so many star parties scattered around the country that it's fairly easy to attend one without having to travel too far or put a strain on your budget.

WHAT TO EXPECT

First and foremost, a star party is a lot of fun, so if you get a chance to attend one (large or small), you should definitely go. It may be intimidating when you first stroll around the grounds and see all the scopes set up, waiting for nightfall. Just remember that you'll be with other astronomy aficionados like yourself, some of whom have recently become enthusiastic about the subject and are also feeling their way around. So don't be shy; introduce yourself and let others know you're just starting out.

You can buy, sell, or trade for stuff at star-party swap meets.

The accumulated knowledge and wisdom at these parties is pretty amazing. Seasoned star-party veterans will be present and are typically very generous with their time and expertise. You'll gain a lot by watching, listening, and picking their brains about astronomy, observing, and gear.

The program varies from party to party, but many have hands-on workshops, observing seminars, and late afternoon or early evening presentations on the hottest topics in astronomy. Occasionally a party may be held near a large observatory, and special tours are often available for those interested in seeing how the pros work.

Another activity common to most star parties is the swap meet. Some are nothing more than notes posted on a bulletin board offering astronomy equipment to trade or sell. At the larger events the swap meet may be a daylong event, complete with tables full of used gear for sale. Of course, it's always buyer beware, but if you're looking for something specific, act quickly — the good stuff tends to rapidly disappear.

The appeal of star parties is worldwide. Each year thousands of astronomy enthusiasts attend the Tainai Star Party near Niigata, Japan.

STAR-PARTY ETIQUETTE

All the rules that apply to any outdoor excursion — be prepared for rain, wind, dust, bugs, humidity, heat, and cold — apply to a star party. But these nighttime events also have a special set of requirements, and if it's your first time, you'll blend right in if you know what they are and follow them.

A star party is a red-light zone — white light is not allowed. Bring a red-light flashlight and carry it with you at all times after dark, though once your eyes adapt to the darkness, you'll be amazed at how well you can see at night without any light at all. Also, watch out for unexpected bursts of white light from your vehicle's interior or trunk light. Either turn them off or cover them with red gel. And no flash photography — you'll definitely get yelled at for trying that!

One of the joys of attending a star party is that everyone present (except those doing astrophotography) will be delighted to have you look through their telescope. Just don't touch or move it without first asking permission. And watch where you walk. Tripping over a cable and crashing into a telescope is considered bad form.

Parking can be an issue. At many star parties you're not allowed to drive your vehicle on or off the observing site after dark. If you need to leave before dawn, park your vehicle away from the observing field with your headlights facing opposite the telescopes (ask an organizer about the best place to park).

Be quiet during the post-sunrise hours. Many folks stay up all night observing, so be careful about rattling around in the camping areas before noon. (Quiet hours are often posted as part of the star-party rules.)

TELESCOPES GALORE

A star party is also a great place to check out equipment. Many of the major telescope manufacturers attend the larger parties to show off their latest and greatest. Even smaller gatherings attract at least a few astronomical-product vendors. Talking one-on-one for a few minutes with the sales reps (many of whom are observers themselves) is an easy way to become better informed about everything from complete telescope systems to filters, dew removers, and other accessories — some of which you won't know you need until you see them!

Most star parties are held at dark-sky sites. However, there are some that take place in or near a city — they're more for the astronomy shopper than the visual observer. Before you go, make sure you read the details of the event so you know what you're getting into.

If you're thinking about acquiring a telescope, a dark-sky star party is your chance to see how well different scopes perform and what type and size might be best for you. Treat the star party as one giant equipment-testing game. Come early and watch people set up: how easy is it to get a big reflector or a portable Go To ready for a night's observing? Under the stars you'll be able to judge for yourself whether it's worth the extra money to get a 10-inch instead of an 8-inch and determine whether an equatorial mount really is for you. And if you find a telescope that's particularly appealing, ask the owner about its good and bad points, the maintenance necessary, and whether it would be good for someone just starting out.

At an observing-based star party, be prepared to stay up all night at least once. When was the last time you did that — and actually enjoyed yourself? Watching a dark, star-studded sky wheel overhead from dusk to dawn is an experience not to be missed.

Whether you bring a telescope, binoculars, or just your eyes, you'll always be made welcome at a star party. Many times I've met partiers from other countries who brought only a big smile and an eagerness to see stars and meet new friends. Party on!

Shopping is another part of the star-party experience. At all of the major parties, vendors of astronomical equipment set up displays of everything from large scopes to tiny accessories. Feel free to ask questions; most vendors are observers, too.

RESOURCES

Get Connected

Now that you know many of my secrets, tips, and tricks, I hope they'll help put you on the path to discovering how fascinating the hobby of astronomy can be. And as I tell many novices, there really is something for everyone to get excited about, because astronomy is so multifaceted that it offers a smorgasbord of fascinating subjects, some of which are sure to pique your interest. So one of the last secrets I'll impart is a very condensed list of astronomy resources — many of which I use all the time — so you don't grow any older looking for the best ones.

Magazines

Perusing the pages of astronomy magazines is a great place to start. These days there are quite a few from which to choose, and your local bookstore should have an assortment. Some large grocery stores and supermarkets also carry them.

Science and technology publications sometimes include astronomy articles describing new discoveries and space technology. Travel magazines will have the occasional article about skywatching; photography magazines sometimes have pieces on simple astrophotography. And don't forget your local library. It may carry astronomy magazines. If not, ask (and get your friends to ask). The library may add a subscription if it perceives that there's an interest.

Sky & Telescope. Established in 1941, *S&T* has a world-renowned reputation for in-depth astronomical news coverage, thought-provoking science articles, and awe-inspiring astrophotography. Easy-to-read charts, constellation descriptions, helpful observing tips, and equipment reviews are just a few of the features packed into each issue.

Sky News. This Canadian magazine of astronomy and stargazing has sky charts tailored for observers north of latitude 45° and includes comprehensive descriptions of sky events, plus easy-to-read news updates.

BBC Sky at Night. A new monthly magazine published in the UK by the BBC, it has the usual observing and science features, and it also comes with a CD full of extras (including Patrick Moore's *Sky at Night* TV program).

Astronomy Now. This British publication has a wide range of feature articles about observing, astronomy, and space science.

Australian Sky & Telescope. As the name implies, this is an Australian version of *S&T*, with a definite Down Under flavor.

Books

Since I've written *Secrets of Stargazing* primarily for novice skywatchers, I'll stick to a handful of the many fine books for beginners that I've either come across or know about.

Patterns in the Sky
Ken Hewitt-White (Sky Publishing)
The night sky becomes familiar territory in this four-seasons guide to the stars and constellations visible from mid-northern latitudes. This is the first book in the *Night Sky: Astronomy for Everyone* series.

NightWatch: A Practical Guide to Viewing the Universe, 4th Edition
Terence Dickinson (Firefly)
An incomparable introduction to the night sky, this book skillfully covers a wide range of observing projects for the naked eye, binoculars, and small telescopes.

The Backyard Stargazer: An Absolute Beginner's Guide to Skywatching with and without a Telescope
Pat Price (Quarry Books)
An informal and informative guide to stargazing, this well-illustrated book touches on a variety of celestial topics and includes dozens of simple and informative observing projects to get you going.

Turn Left at Orion
Guy Consolmagno and Dan M. Davis
(Cambridge University Press)
This excellent observing guide for those with a small telescope includes 100 deep-sky sights with sketches of their appearance in the eyepiece.

Binocular Highlights
Gary Seronik (Sky Publishing)
This book is a reprint of 74 Binocular Highlight columns from *Sky & Telescope* magazine. It contains nearly 100 objects: most are easy to spot, but some are challenging.

The Monthly Sky Guide, 7th Edition
Ian Ridpath and Wil Tirion
(Cambridge University Press)
An incisive month-by-month description of the night sky, this guide includes excellent charts, and close-ups of interesting star regions. It's fully revised and updated for planet positions and eclipses up to the end of the year 2011.

365 Starry Nights
Chet Raymo (Simon & Schuster)
A fun and informative guide for every night of the year, it contains funky charts, whimsical drawings, and easy-to-read maps. It's a hard-to-put-down introductory stargazing companion.

Observer's Handbook
Royal Astronomical Society of Canada
I can't call this annual a "beginners" book, but it's an essential yearly purchase for all serious observers. The *Handbook* contains Sun, Moon, and planet data for the current year plus a wealth of what-to-see observing information. It can be intimidating when you first pick it up, but it'll soon become an essential part of your observing kit.

Star Atlases

Even if you're just starting out, you need a star atlas. Merely owning one will stimulate your desire to go out and try to observe some of the sights marked on the star maps. There are plenty of atlases on the market, so I've tried to select a few that will serve you well as you progress from novice to experienced stargazer.

***Sky & Telescope*'s Star Wheel and *Night Sky* Star Wheel**
(Sky Publishing)
Think of these planispheres as simplified star atlases that will show you which constellations are visible on any date and time. *S&T*'s Star Wheel is available for different latitudes; the smaller *Night Sky* Star Wheel is designed for use in light-polluted skies.

***Sky & Telescope*'s Pocket Sky Atlas**
Roger W. Sinnott (Sky Publishing)
This portable 6- by 9-inch star atlas is spiralbound, folds flat, and is perfect for field use. It has 80 charts, more than 30,000 stars to magnitude 7.6, and 1,500 deep-sky objects. It's new and is already a must-have!

Bright Star Atlas
Wil Tirion and Brian Skiff (Willmann-Bell)
Covering the sky in 10 charts, the maps contain nearly 9,000 stars with visual magnitudes as dim as 6.5 and include more than 600 bright deep-sky objects.

The Cambridge Star Atlas, 3rd Edition
Wil Tirion (Cambridge University Press)
The 20 charts in this basic atlas for stargazers show stars down to magnitude 6.5 and plot 900 deep-sky objects.

Sky Atlas 2000, Second Edition
Wil Tirion and Roger W. Sinnott
(Sky Publishing)
This is the atlas to have if you're serious about star-hopping. It plots more than 85,000 stars to magnitude 8.5 as well as 2,700 deep-sky objects — enough sights to provide for a lifetime of observing.

Software

At their most basic, desktop planetarium programs are an electronic version of a planisphere. Yet that's not really a fair description because they do so much more. You can use them to view a representation of the night sky as it appears in any direction at any time in the past or future, from anywhere on Earth. Moreover, they can speed up time to show the motions of the Moon, stars, and planets. Recommending one over another is difficult; sometimes it boils down to personal preference. So with that in mind, here are two that I like.

MegaStar 5
(Willmann-Bell)
This is a planetarium program with more than 208,000 deep-sky objects contained in three different star atlases that show more than 15 million stars. Users can customize charts for their individual requirements (thank goodness), but note that it doesn't run on a Mac. The program will also interface with some Go To telescopes.

Starry Night
(Imaginova Corp.)
Imaginova has produced a suite of desktop planetarium programs, ranging from an inexpensive but nice "starter" program to one that uses a full-color all-sky CCD mosaic that realistically depicts the night sky. It's expensive, but no matter what the weather you'll always have stars to look at! When the sky is clear, you can use the program to interface with some Go To scopes.

World Wide Web

The World Wide Web should be called the Universal Web because it's so vast that I'm not sure it has an end. A quick online query using the words "astronomy resources" turned up more than 75,000,000 listings! So I've simply noted some of the major sites and places I like to frequent, and have done so in alphabetical order.

(In this book I've occasionally mentioned different companies and the various products they sell. I'm reluctant to promote one and not another, so please use your favorite search engine to find the websites that offer whatever product you're looking for.)

Abrams Planetarium's Starline — download a 3-minute podcast that describes the whereabouts of the planets in the night sky.
www.pa.msu.edu/abrams/starline/index.html

Accuweather — weather-forecast site with podcast and PDA downloads
Accuweather.com

APOD (Astronomy Picture of the Day) — has superb images and explanations.
http://antwrp.gsfc.nasa.gov/apod/astropix.html

Clear Sky Clock — shows at a glance 48-hour-long periods of the expected cloud cover, transparency, and seeing for many North American and Canadian sites.
http://cleardarksky.com/csk

International Dark-Sky Association — if light pollution is affecting your observing, take action!
www.darksky.org

Heavens Above — contains real-time satellite information and visibility predictions (including the International Space Station and Space Shuttle).
Heavens-above.com

Jack Horkheimer: Star Gazer — a weekly TV series on naked-eye astronomy.
www.jackstargazer.com

NOAA National Climatic Data Center — for weather predictions.
http://lwf.ncdc.noaa.gov/oa/ncdc.html

Spaceweather.com — what's up in space.
www.spaceweather.com/

Sky Publishing — the publisher of *Sky & Telescope* magazine has a new website that includes observing articles, astronomy news, product information, podcasts, blogs, an interactive star chart, and other items of interest for beginners and experienced observers.
SkyandTelescope.com

StarDate Online — listen to short radio programs about the sky.
http://stardate.org

US Naval Observatory — the data-services page includes Sun and Moon rise and set times, Moon phases, eclipses, seasons, positions of solar-system objects, and other data.
http://aa.usno.navy.mil

Online chat groups are another way to plug into topics that cover just about anything astronomical. Some cater to very specific topics, such as a particular brand of telescope, mount, or eyepiece. On the one hand, these groups can provide useful information and often have members with knowledge and first-hand experience that they're willing to share. On the other hand, not all of the advice is sound, and sometimes the chats degenerate into boorish behavior and "shouting matches." I suggest, at least initially, sticking to moderated forums. Use your favorite search engine to find a group that discusses a topic you're interested in, and then check out their FAQ (Frequently Asked Questions) section.

Star Parties

As I mentioned earlier, attending a star party is an invaluable way to learn and connect with others who share your interest in the night sky. So many astronomy clubs hoid these events that I simply can't include them all. I've listed a few of the major ones worldwide (listed in alphabetical order; I'm sorry if I missed yours), but I'll start with one website that can help you find a star party in your area.

Sky Publishing — the event-calendar page contains lots of options for locating a regional or local star party, but the data is only as good as the information received by Sky Publishing.
SkyandTelescope.com/community/calendar

European Astrofest — no observing, but offers plenty of information about astronomy and space (in London, England).
www.astronomynow.com/astrofest

Northeast Astronomy Forum & Telescope Show — America's premier astronomy expo (in Suffern, New York).
www.rocklandastronomy.com/neaf.htm

Oregon Star Party — has become one of the largest gatherings in the US (in the mountains of central Oregon).
www.oregonstarparty.org

Queensland Astrofest — a gathering of amateurs in northeastern Australia (northwest of Brisbane, Queensland).
www.qldastrofest.org.au/

RTMC Astronomy Expo — a popular forum for equipment and observing (northeast of Riverside in the San Bernardino mountains in California).
www.rtmcastronomyexpo.org

South Pacific Star Party — one of Australia's major star parties (west of Sydney, New South Wales).
www.asnsw.com/spsp/spsp.htm

Starfest — Canada's largest annual observing convention and star party (outside of Toronto, Ontario).
www.nyaa-starfest.com/starfest

Stellafane — a major gathering of amateur telescope makers (Vermont).
www.stellafane.com

Table Mountain Star Party — a large observing party 6,500 feet high on Table Mountain (central Washington State).
www.tmspa.com/

Tainai Star Party — thought to be the world's largest annual gathering of sky enthusiasts (Kurokawa Village, north of Tokyo, Japan).
www.tainai.jp/tainai/indexe.html

Texas Star Party — one of the largest gatherings in the US (southeast of El Paso).
www.texasstarparty.org

Winter Star Party — a favorite for northern observers because it's held in the Florida Keys in winter!
www.scas.org/wsp2004.html

Astronomy Clubs

The importance of joining an astronomy club has already been covered, but I'll repeat it here. Your local club offers resources, knowledge, and experience, and you'll make lasting friendships that are priceless. There are far too many clubs to list, so instead I'll mention two websites that'll help you locate a club near you.

Sky Publishing has an extensive (searchable) list on its website.
SkyandTelescope.com/community/organizations

The **Association of Lunar and Planetary Observers** has an incredible page on its website that lists hundreds of clubs in the US and around the world (by country).
www.lpl.arizona.edu/~rhill/alpo/clublinks.html

In "Finding a Club" on page 79, I mentioned that one way you can do just that is to contact your local planetarium or science center. While you probably know of a nearby major facility, you might not be aware of a small planetarium lurking in your neighborhood. So here are two more sites to help in your search for fellow night-sky enthusiasts.

On the Clubs & Organizations page of the **Sky Publishing** website, enter your country and state/province to get the largest selection; enter your city to narrow your search.
SkyandTelescope.com/community/organizations

This page on the **International Planetarium Society**'s website will send you to an extensive PDF list of planetariums worldwide.
www.ips-planetarium.org/atw/ips-around.html

APPENDIX 1

A Bag of Tricks

It's only common sense to pack a "bag of tricks" for you and your telescope. A few carefully selected potential necessities, stashed in a bag that goes wherever your scope goes, could save your observing adventure if something breaks or stops working. Here's a list of the contents of my "Don't leave home without it" bag of tricks. You'll notice that it doesn't include observing necessities like star charts or eyepieces; all that stuff is a given. My list is in no particular order (though I have tried to group similar items). I'm not suggesting you take everything I do, and depending on your scope and situation, there may be other items that you feel should be included in *your* bag of tricks that I haven't listed. Feel free to improvise.

Not everything I describe in this section is shown here. Some of the necessities are part of a kit that lives in my vehicle and always goes with me when I'm on the road.

A card with emergency contacts and phone numbers, a cell phone, and a few dollars for "just in case." Also throw in a couple of quarters on the off chance that you end up having to use a pay phone. (You do remember those, don't you?)

Duct tape. The astronauts don't leave home without it; why should you? Don't forget electrical tape for emergency electrical repairs and wire wrapping.

Bungee cords are essential if you're trying to tie down rowdy equipment or flapping scope covers, or if you've got so much stuff that your car's trunk won't close! While we're on the subject of keeping things under control, bring along some fabric-covered hair scrunchies (they're like mini–bungee cords and can be used in tight spots) and some rubber bands and/or twist ties (good for restraining unruly and overly long wires and cables).

Paper clips secure papers and can be used to mark pages. They're also good for bending into strange-shaped tools for repairing things you never thought would break or as an emergency replacement (albeit not a very secure one) for a lost nut-and-bolt pair.

A pocketknife is useful, but even better is a multi-purpose tool, which looks like a jazzed-up Swiss army knife. It has more widgets and gadgets than you could possibly use in a lifetime but will need as soon as you don't take it with you.

If possible I have a tarpaulin or ground sheet to put under my scope — this makes it easier to find dropped items. Even with ground cover, a heavy-duty magnet is useful for attracting those tiny metal screws, nuts, and bolts that inevitably plunge to the ground. In the same vein, I've often been thankful that I carry tweezers for picking minute things out of the tiny crevices.

A couple of clamps will be handy on windy nights for keeping papers and charts securely fastened to your little observing table. Speaking of fastening, Velcro tabs and Velcro tape have more uses than you can imagine. As for tables, a few small blinky red lights are useful as anti-collision markers around the base of your table, telescope, and step stool. Just don't go overboard and use too many — you'll annoy your neighbors.

Carry either an extra red-light flashlight or, if that's a bit much, have extra batteries for your flashlight. If your telescope needs battery power, bring extras for it, too.

Are you recording your observations? If you're doing it the "old-fashioned" way, take along a spare pencil and maybe a small extra notebook. If you're using a camera, have extra batteries and an extra memory card handy.

Under the heading of miscellaneous stuff, I suggest that you include a hand magnifier (or a loupe) for reading charts when your eyes are fatigued and not performing optimally; a small spirit or bubble level to help level your telescope (if the scope doesn't have one built in); a Lenspen for quick spot-cleaning of eyepieces; and a set of jeweler's screwdrivers. Bring a towel or two (one large and one small) for warmth, padding, ground cover, or spill cleanup. Towels are almost as versatile as duct tape!

If you're serious about getting into the country to a dark-sky site, bring a GPS (in addition to your maps). If you don't, at least have a compass with you.

In my opinion, you should have a space blanket in your observing kit and in your car that'll keep you warm and dry in an emergency. It can also act as a cover for your telescope to keep it cool, deter dust, or even temporarily keep it dry if a rainstorm catches you off guard.

Finally, here's an unusual one. Hand sanitizer is good not only for dissolving goopy residue on your hands, but it can also be used as an emergency fire starter — if you purchase the type of hand cleanser that has alcohol as its main ingredient. If you're desperate enough to use it in this manner, be very careful (you don't want to add to your woes). Of course this tip is useless if you don't also carry matches.

DON'T OVERLOOK THE OBVIOUS

If you wear glasses or contact lenses, remember to pack spares, particularly if you need them for viewing. Keep them clean. Dust and dirt on glasses can degrade your view, and contacts can accumulate protein buildup that may make their use uncomfortable.

APPENDIX 2

Becky's Urban Favorites

Although this isn't a "what to look at" observing book, I felt that including a short list of sights easily seen by urban observers was appropriate. I've included the coordinates (RA and Dec.) of the 15 deep-sky objects so those with Go To scopes can simply enter the object's celestial address, let the scope do its thing, and then enjoy the view. If you don't have a Go To, you'll need a basic star atlas to locate them.

The Moon. Watching mountain shadows move across crater floors and seeing isolated peaks that seem to "float" above the lunar limb when they are first illuminated is particularly hypnotic.

The Planets. I never tire of hunting **Venus** in the daytime or finding **Mercury** when it's visible. **Jupiter** and its moons are always fascinating. However, **Saturn** with its gorgeous rings was my first true love.

The Sun. When it's freckled with sunspots, it's fun to watch and sketch their daily progress and changes. Just remember to use a solar filter.

Satellites. Spotting them as they pass overhead is always a thrill.

Now for some of my favorite, urban accessible, deep-sky sights, most of which can be observed with binoculars or a small telescope.

CELESTIAL COORDINATES

Everything has an address. On Earth we have longitude and latitude. In the sky, latitude corresponds to *declination* (dec.). This denotes, in degrees, how far north or south of the celestial equator an object is in the sky. Polaris is almost +90°. (The celestial equator is just an imaginary extension of Earth's equator into space).

The celestial equivalent to geographic longitude is *right ascension* (RA). The scale increases in hours eastward from 0 to 24, and Earth's rotation causes the sky to move through one hour of right ascension in just about an hour of time.

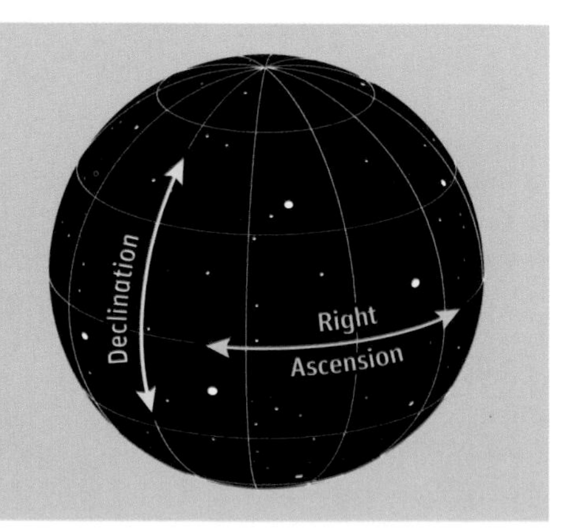

Double Cluster

WHAT'S IN A NAME?

If you look through any catalog of celestial objects, you'll notice that many have nicknames: E.T. Cluster, Beehive, Coathanger. But what about M8 or NGC 457? The 18th-century French astronomer Charles Messier catalogued some of the sky's brightest star clusters and nebulas in his search for comets. They're now known as M objects in his honor. In 1888 the Danish-Irish astronomer J. L. E. Dreyer published a New General Catalogue containing 7,840 deep-sky objects; the sights are now known as NGC objects.

Popular name	Constellation	RA	Dec.	Mag.	Comments
Andromeda Galaxy	Andromeda	$00^h 42.7^m$	+41° 16′	3.5	M31
E.T. Cluster	Cassiopeia	$01^h 19.6^m$	+58° 17′	6.4	NGC 457
Double Cluster	Perseus	$02^h 19.0^m$	+57° 08′	4.3, 4.4	NGC 869, NGC 884
Pleiades	Taurus	$03^h 47^m$	+24° 07′	1.6	M45, Seven Sisters
Great Orion Nebula	Orion	$05^h 35.4^m$	-05° 27′	4.0	M42
Beehive Cluster	Cancer	$08^h 40^m$	+19° 40′	3.1	M44
Great Cluster in Hercules	Hercules	$16^h 41.7^m$	+36° 28′	5.9	M13
False Comet	Scorpius	$16^h 54.2^m$	-41° 50′	2.6	NGC 6231, rich open cluster
Lagoon Nebula	Sagittarius	$18^h 03.9^m$	-24° 23′	5.8	M8
Sagittarius Cluster	Sagittarius	$18^h 36.4^m$	-23° 54′	5.2	M22, Great globular
Wild Duck Cluster	Scutum	$18^h 51.1^m$	-06° 16′	6.3	M11
Coathanger	Vulpecula	$19^h 25.4^m$	+20° 11′	n/a	Brocchi's Cluster, Collinder 399
Albireo	Cygnus	$19^h 30.7^m$	+27° 58′	3.1, 5.1	Beautiful double star
Omicron[1] Cygni	Cygnus	$20^h 13.6^m$	+46° 44′	3.8, 4.8	Turquoise & orange pair
Garnet Star	Cepheus	$21^h 43.5^m$	+58° 47′	3 - 5	Mu Cephei, an easy variable

Glossary

altazimuth (altaz): A simple telescope mount with altitude (up/down) and azimuth (left/right) motions.

angular size/distance: The apparent size of an object in the sky or the distance between two objects, measured as an angle. The Sun and Moon are ½° across; your fist held at arm's length spans 10°; from the horizon to the zenith is 90°.

aperture: The diameter of a telescope's main lens or mirror, expressed in millimeters or inches.

asterism: A pattern of stars that isn't a whole constellation. The Big Dipper is an asterism in the constellation Ursa Major.

Astronomy Day. This internationally recognized event occurs annually between mid-April and mid-May on a Saturday near or before first-quarter Moon.

averted vision: Observing an object by looking slightly to one side of where it is. This can be very helpful when you're trying to see a faint nebula or galaxy that's invisible when you stare directly at it.

Barlow lens: A lens placed between the eyepiece and the focusing tube (or star diagonal) that doubles or triples the magnification of the eyepiece with which it's used.

catadioptric (compound): A telescope that uses both a lens and mirrors to gather light.

celestial coordinates: A grid system used to position objects in the sky. Declination and Right Ascension are the celestial equivalents of latitude and longitude.

celestial equator: An imaginary line extended out from Earth's equator that divides the sky in half.

collimation: The process of aligning the center of your telescope's primary mirror or lens so that it aims at the center of the scope's eyepiece. This is done most often on reflectors.

constellation: A pattern of stars used to organize a segment of the sky. The word refers both to the 88 official constellations, which define sections of the sky, and to the shape of the mythical star pattern within the borders of each sky segment.

dark adaptation: The ability of our eyes to adjust to low levels of light. It takes approximately 20 to 30 minutes for the eye to become reasonably adapted to the dark.

declination (dec.): The celestial equivalent of latitude. It denotes (in degrees) how far north or south of the celestial equator an object is.

deep-sky object: Any celestial object located beyond the solar system, including stars, star clusters, nebulas, and galaxies.

Dobsonian (Dob): A Newtonian reflector on a simple wooden altazimuth mount. Amateur astronomer John Dobson invented and popularized this type of scope.

eyepiece: The part of a telescope that you look into.

field of view: The amount of sky that you see when you look through a telescope or binoculars. Generally the higher the magnification, the tinier the field of view.

finderscope (finder): A low-power, wide-field mini-telescope that's attached to the main instrument and is used to aim the telescope at a celestial object. (See unit-power finder.)

focal length: The distance from the telescope's objective (its primary lens or mirror) to the point where an image comes into focus. This distance is usually expressed in millimeters.

focal ratio (f/number): A telescope's focal length divided by its aperture. For example, a telescope with a 2000-mm focal length and a 200-mm-diameter mirror has a focal ratio of f/10.

Go To telescope: A computerized scope (the

computer is usually located in the mount) that can find, and track, thousands of preselected celestial objects.

light pollution: The upward glow from artificial light that adversely affects the night sky. If you live in a city, light pollution greatly reduces the number of stars you can see.

magnification (power): The apparent increase in size of an object when viewed through a telescope. If something is magnified 50 times, this is written as "50×" and is spoken as "50 power."

magnitude: The measurement of the brightness of a celestial object. The higher the numerical value of the magnitude, the fainter the object.

Messier object: One of 110 objects in a catalog begun in 1758 by French comet hunter Charles Messier. All are visible in a small telescope.

Milky Way: The broad, faintly glowing band of light stretching across the night sky; it's particularly prominent in summer but only when seen from a dark sky site. The glow comes from billions of stars too faint to be seen individually. It's also the name we give to the spiral galaxy in which our Sun is located.

mount: The device on which a telescope sits. Some are motorized to aid in the tracking of stars; others are motorized *and* have an onboard computer.

NGC (*New General Catalogue*): A catalog of 7,840 deep-sky objects originally published in 1888 by Danish-Irish astronomer J. L. E. Dreyer. Today it forms the core database for all computerized telescopes.

objective: A telescope's main light-gathering mirror or lens.

reflector: A telescope that gathers light with a mirror.

refractor: A telescope that gathers light with a lens.

right ascension (RA): The celestial equivalent of longitude. The sky moves through one hour of right ascension in just about an hour.

seeing: A measure of the atmosphere's stability. When the stars are twinkling like mad and you can't get them to stay focused in your scope, the atmosphere is turbulent and astronomers say the seeing is bad.

skyglow: The weird, rosy-pink-colored light that looks like smeared cotton candy hovering overhead at night. It's caused by light reflecting off tiny airborne particles of dust, debris, and moisture.

star diagonal: A mirror or prism in an elbow-shaped housing that attaches to the focuser of a refractor or compound telescope (a reflector doesn't need one).

star-hopping: A technique used by amateur astronomers to move from a bright star to a series of dimmer stars with the ultimate goal of finding a target that's more than likely below naked-eye visibility. A Go To telescope avoids the need to star-hop.

star party: A gathering of astronomy enthusiasts, usually in a dark-sky location. Astronomy clubs often hold urban star parties to introduce astronomy to the public.

transparency: A measure of the atmosphere's clarity. It's ironic that crystal clear nights with superb transparency sometimes have poor seeing.

twilight: The time before sunrise or after sunset when the sky is not fully dark.

unit-power finder: A device that attaches to the main instrument, shows the sky as it appears to your eye without magnification, and is used to aim the main scope at a distant celestial sight. Some versions project a red dot, circle, or crosshairs onto a viewing window.

zenith: The point in the sky that's directly overhead.

Index

Page numbers in **boldface** mean that the subject appears in a photograph, chart, or illustration.

Acknowledgments

Portions of "In Your Own Words" in Chapter 3 first appeared in *Night Sky* magazine, July/August 2005, in the article "Dear Sky Diary."

Image Credits

INTRODUCTION **vi** Becky Ramotowski

CHAPTER 1 **p2** Becky Ramotowski; **p3** Rik Davis; **p4** [top] David Regen, [bottom] Becky Ramotowski; **p5** [top] Becky Ramotowski, [middle] Damian Peach, [bottom] Gary Seronik; **p6** [top] Becky Ramotowski, [bottom] SkyShed; **p7** Sky Publishing: Craig Michael Utter; **p8** Becky Ramotowski; **p9** [top] Sky Publishing: Tony Flanders (×3), [bottom] Becky Ramotowski.

CHAPTER 2 **p10** Sky Publishing: Craig Michael Utter; **p11** Sky Publishing: Craig Michael Utter; **p12** Sky Publishing: Gregg Dinderman; **p13** Becky Ramotowski; **p14-15** [top] Sky Publishing: Gregg Dinderman (×3); **p14-15** [bottom] Sky Publishing: Craig Michael Utter (×4); **p16** [top] Sky Publishing: Edwin L. Aguirre and Imelda B Joson, [bottom] Sky Publishing: Craig Michael Utter; **p17** [top] Akira Fujii, [bottom] Sky Publishing: Craig Michael Utter; **p18** Sky Publishing: Craig Michael Utter; **p19** Lee Labuschagne; **p20** Sky Publishing: Gregg Dinderman; **p21** Sky Publishing: Craig Michael Utter (×2); **p22** Akira Fujii; **p23** Sky Publishing: Craig Michael Utter (×4); **p24** [top left, bottom] Sky Publishing: Craig Michael Utter, [top right] Sky Publishing; **p25** Sky Publishing (×2); **p26** Sky Publishing: Craig Michael Utter; **p27** Sky Publishing: Craig Michael Utter (×5); **p28** [top] Sky Publishing: Craig Michael Utter, [bottom] Sky Publishing: Gregg Dinderman; **p29** Becky Ramotowski; **p30** Sky Publishing: Craig Michael Utter (×2); **p31** Sky Publishing: Craig Michael Utter (×2); **p32** [top] Sky Publishing: Sean Walker (×3), [bottom] Sky Publishing: Paul Deans; **p33** [top] Sky Publishing: Craig Michael Utter (×2), [bottom] Becky Ramotowski; **p34** [top] Sky Publishing: Craig Michael Utter, [bottom] Kirk Pu'uohau-Pummill / Gemini Observatory; **p35** Sky Publishing: Craig Michael Utter (×2); **p36** Akira Fujii; **p37** [top] Sky Publishing: Gregg Dinderman, [bottom] Sky Publishing: Craig Michael Utter; **p38** Sky Publishing: Gregg Dinderman; **p39** Sky Publishing: Craig Michael Utter (×3); **p40** Sky Publishing: Craig Michael Utter (×3); **p41** [top] Becky Ramotowski, [bottom] Sky Publishing: Craig Michael Utter.

CHAPTER 3 **p42** Akira Fujii; **p44** [top] Sky Publishing: Richard Tresch Fienberg, [bottom] Sky Publishing: Craig Michael Utter; **p45** Becky Ramotowski; **p46** [top] Becky Ramotowski, [bottom] Jack Newton; **p47** [top] Gary Seronik, [bottom] Sky Publishing: Dennis di Cicco; **p48** National Weather Service / NOAA; **p49** Becky Ramotowski (×4); **p50** Alan Dyer; **p51** Sky Publishing: Richard Tresch Fienberg; **p52** Becky Ramotowski (×2); **p53** Sky Publishing: Craig Michael Utter; **p54** Sky Publishing: Craig Michael Utter; **p55** [top left] Sky Publishing: Stuart J. Goldman, [middle & bottom] Becky Ramotowski; **p56** Sky Publishing: Craig Michael Utter (×3); **p57** Sky Publishing: Craig Michael Utter (×2); **p58** [top] Sky Publishing: Craig Michael Utter, [bottom] Sky Publishing: Dennis di Cicco (×2); **p59** [left] Sky Publishing: Craig Michael Utter, [right] Alan Dyer; **p60-61** Sky Publishing: Tony Flanders (×9); **p62** Sky Publishing: Craig Michael Utter; **p63** [top] Akira Fujii, [middle] Sky Publishing: Gary Seronik; **p64** Akira Fujii; **p65** Sky Publishing: Craig Michael Utter (×2); **p66-67** Sky Publishing: Craig Michael Utter (×4); **p68** Becky Ramotowski; **p69** Becky Ramotowski (×3); **p70** Sky Publishing: Craig Michael Utter; **p71** Becky Ramotowski; **p72** Sky Publishing: Craig Michael Utter; **p73** Akira Fujii; **p74** Akira Fujii; **p75** [top] Akira Fujii, [middle] Russell Croman.

CHAPTER 4 **p76** Tony & Daphne Hallas; **p78** [top] Sky Publishing: Dennis di Cicco; [bottom] Sky Publishing: Richard Tresch Fienberg; **p79** Becky Ramotowski; **p80** [top left] Jim Failes, [top right] Steve McGough, [bottom] Cradle of Aviation Museum; **p81** Sky Publishing: Richard Tresch Fienberg; **p82** [top] Sky Publishing: Richard Tresch Fienberg, [middle] Sky Publishing: Gary Seronik, [bottom] Sky Publishing: Dennis di Cicco; **p83** Sky Publishing: Paul Deans.

CHAPTER 5 **p84** David Dunn.

APPENDICES **p90** Becky Ramotowski; **p92** Sky Publishing; **p93** Akira Fujii.